The Nation's #1 Educational Publisher

McGraw·Hill
Learning Materials
SPECTRUM
GEOGRAPHY
WORLD

Grade 6

Authors

James F. Marran
Social Studies Chair Emeritus
New Trier Township High School
Winnetka, Illinois

Cathy L. Salter
Geography Teacher
Educational Consultant
Hartsburg, Missouri

McGraw·Hill
Learning Materials

8787 Orion Place
Columbus, OH 43240-4027

The McGraw·Hill Companies

EAN

9 781577 681564

90000

Program Reviewers

Bonny Berryman
Eighth Grade Social Studies Teacher
Ramstad Middle School
Minot, North Dakota

Grace Foraker
Fourth Grade Teacher
B. B. Owen Elementary School
Lewisville Independent School District
Lewisville, Texas

Wendy M. Fries
Teacher/Visual and Performing Arts Specialist
Kings River Union School District
Tulare County, California

Maureen Maroney
Teacher
Horace Greeley I. S. 10 Queens
District 30
New York City, New York

Geraldeen Rude
Elementary Social Studies Teacher
1993 North Dakota Teacher of the Year
Minot Public Schools
Minot, North Dakota

Photo Credits

McGraw-Hill
Consumer Products

A Division of The McGraw·Hill Companies

Table of Contents

Modern Europe and the Rise of the Soviet Union

Middle East and North Africa

Africa South of the Sahara

South Asia

Lesson 1

When the World Was Frozen

As you read about Earth during the Ice Age, think about how the climate affected the way people lived.

About 18,000 years ago Earth looked much different than it does today. Imagine what would happen if today's astronauts could travel back to that time, which is known as the **Ice Age.** From space, they would see **glaciers,** or large sheets of ice, blanketing much of North America and parts of Europe and Asia. There was even a glacier on Hawaii.

In addition to seeing large patches of white glaciers, the astronauts would notice wide bands of brown in the northern hemisphere. Because of the cold climate, much of the land just south of the glaciers was treeless and dry. They would see green stripes of forests much farther south than they are today.

To the time-traveling astronauts, even the outlines of the continents would seem strange. So much of the ocean's water was trapped in glaciers that sea levels were hundreds of feet lower than they are now. Land that is underwater today was above water then. For example, a wide land bridge linked North America with Asia.

This Ice Age shelter was made from mammoth tusks and bones.

In dark caves, Ice Age people painted and carved images of the animals they hunted.

The people who lived during the Ice Age had to cope with weather that was much colder than it is today. To survive the bitterly cold winters, they learned to build fires and find shelter in warm caves. During the warmer summers, they lived in tents made of bones and hides.

Ice Age people hunted reindeer, woolly mammoths, bison, and other large animals for food, hides for warmth, and bones. They used tools carved from reindeer bones to sew warm clothes from animal skins. Wooden spears with sharp stone tips helped hunters bring down their prey. People also fished with carved hooks and harpoons. Fruits, nuts, and roots rounded out the Ice Age diet.

People of the Ice Age followed large herds of animals as they migrated north during the summer and south during the winter. When herds grew scarce in Eurasia, some groups crossed the land bridge between East Asia and North America looking for new hunting grounds. These Ice Age hunters were the first settlers of the Americas.

When the world was frozen people had to live in a cold, harsh climate. The climate affected the types of shelters they lived in, the clothes they wore, and the food they ate. Despite the harsh conditions, people of the Ice Age settled the land.

Lesson 1

MAP SKILLS Using Special-Purpose Maps

The two maps shown below are **special-purpose maps,** which focus on a specific topic. One map shows the moisture, ice extent, and forests of North America during the Ice Age. The other map shows the same features for the present day. By looking at the shading on each map, you can see how different North America was during the Ice Age.

Map 1:
Ice Age Moisture and Vegetation

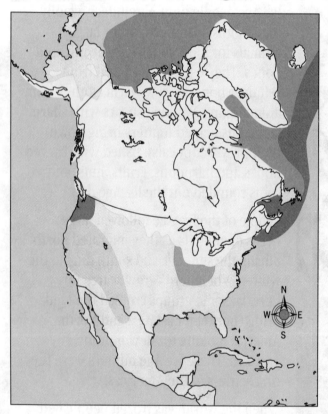

Map 2:
Present-Day Moisture and Vegetation

Map Key

Moisture

Drier than today

Same as today

Wetter than today

Ice Extent

Ice sheets

Winter-only sea ice

Year-round sea ice

Vegetation (forests)

Spruce-rich forest

Oak-rich forest

A. **Study the climate and vegetation map of the Ice Age and answer the following questions.**

1. What was the general moisture of North America like during the Ice Age?

   Same as today

2. What was the general moisture of present-day United States like during the Ice Age?

   Same as today

3. What was the primary vegetation during the Ice Age?

   Spruce-rich forest

4. What part of present-day North America was covered with ice sheets?

   North-East U.S. + Canada

B. **Study the present-day climate and vegetation map to answer the following questions.**

1. What is the primary vegetation in the United States on the present-day map?

   Oak-rich forest

2. Where is the spruce-rich forest today?

   Canada + Alaska

C. **Use both climate and vegetation maps to answer the following questions.**

1. How did the ice affect present-day vegetation?

   The ice became Spruce-rich forests.
   Oak rich in east U.S.

2. How would you describe the change in the patterns of forests between the Ice Age period and present-day? Explain the role melted glacial ice played in the change in vegetation.

   The glacial made the soil richer for lots
   of Spruce-rich forests.
   It was wetter is someplaces and drier
   in someplaces.

Lesson 1

ACTIVITY
Describe how the physical environment affects the way people live.

The Impact of the Environment

The **BIG** Geographic Question

How does the environment affect the way people live?

From the article you learned how Ice Age people coped with life in a cold climate. The map skills lesson showed you the climate's effect on vegetation during the Ice Age and the present day. Find out about how the physical environment influences people's lives.

A. Using information from the article, think about characteristics of the Ice Age environment. Complete the following.

1. Describe the land that was not covered by glaciers during the Ice Age and describe where this land was located.

   South of glaciers were treeless and dry.

2. What physical conditions did people who lived during the Ice Age encounter and how did they adapt to those conditions?

   The built fires and lived inside caves.

3. What were the movement patterns of people during the Ice Age, and what factors influenced them?

   They traveled with the migrateing animals, moving north or south. They moved to other hunting grounds when herds grew scarce.

B. Using the data from the article and the map skills lesson, write three generalizations about how the physical environment influenced the shelter, clothing, and food of Ice Age people. Then write three generalizations about how the environment affects people today.

	Ice Age	Present Day
Shelter	Caves Tents of Bones & Hides	Apartments Homes Foster Homes
Clothing	Mamoth Reindeer and Bison hides	Tee-Shirts Jeans Baseball caps
Food	Friuts Nuts Roots	Pizza Mac & Cheese Steak

C. From the chart you can see how the physical environment affects the way people live. But the reverse is true, too—people affect the environment in which they live. Which of the two previous statements do you think most appropriately applies to the Ice Age and which to the present? Explain your answer.

The Ice Age applies to #1 because if it snowed or was hot they needed to find new shelter. Now people build buildings + houses and the animals have to move.

Lesson 2
Angry Waves

As you read about a tsunami that occurred in ancient times, think about how water can affect life on Earth today.

With awesome force, a volcano on the island of Strongyle (STRAWN-juh-lee) erupted in 1450 B.C. The terrible blast shot lava more than 20 miles into the sky and ripped apart an island in the Aegean Sea. As Strongyle's volcano

How a Tsunami Is Created

1 When an underwater earthquake occurs, the sea floor moves up or down.

2 This movement displaces, or pushes up, large amounts of water and sends waves fanning out from that point. These waves create ripples on the surface of the water. You can see the same process on a smaller scale. Imagine you are riding in a car holding a glass of water. All of a sudden the car hits a bump in the road and the water comes splashing over the top of the glass. The movement happening beneath the cup as the car hits the bump causes the water to splash. The ripples of a tsunami are similar to the ones you see when you throw a stone in a pond, only stronger.

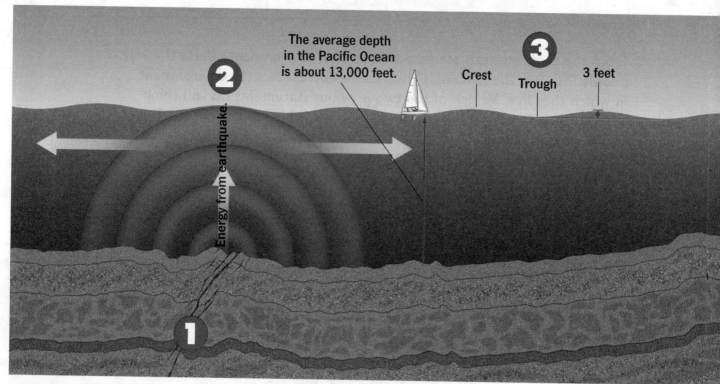

The average depth in the Pacific Ocean is about 13,000 feet.

Energy from earthquake.

Crest Trough 3 feet

disintegrated into an enormous crater, it created a crushing wave more than 600 feet tall. Three great waves, each the height of a 20-story building, pounded the island. The roaring water stormed inland for miles, rising hundreds of feet up the sides of mountains.

The wave that destroyed the people of Strongyle in the fifteenth century B.C. was a type of wave called a **tsunami** (soo-NAH-mee). *Tsunami* is a Japanese word meaning "harbor wave." Tsunamis are caused by earthquakes, volcanic eruptions, or underwater landslides. Though they are relatively rare events, occurring about once every ten years, they have killed about 100,000 people over the centuries. Tsunamis are extremely destructive to land and people in their paths.

3 A tsunami's crest—the top of the waves—is usually about three feet high. Although tsunamis can move at speeds of about 500 nautical miles per hour, out on the open sea they can pass under a ship without sailors noticing them.

4 As a tsunami approaches land its threat grows. As the tsunami passes over more shallow water, friction from the sea floor causes it to slow down. Because it still has the same volume of water, the waves pile up, and the tsunami's crest rises higher and higher.

5 By the time the tsunami reaches shallow water close to shore, the height of the crest can range from a few feet to hundreds of feet.

Lesson 2
MAP SKILLS Using a Legend to Identify Ocean Depth

Maps that show ocean depth can help us find out where a tsunami might form and how dangerous it might become. Most tsunamis originate in the trenches, or the deepest parts, of the Pacific Ocean. Scientists can measure how fast a tsunami is moving based on the depth of the sea through which it travels.

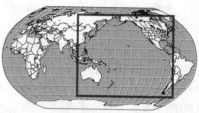

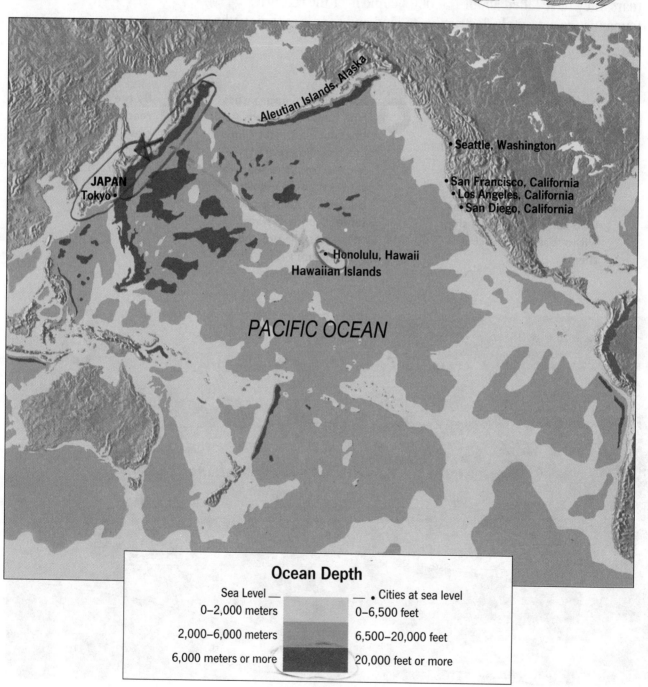

Ocean Depth

Sea Level		· Cities at sea level
0–2,000 meters		0–6,500 feet
2,000–6,000 meters		6,500–20,000 feet
6,000 meters or more		20,000 feet or more

A. Maps show ocean depth and area at sea level. Study the map and legend and complete the following.

1. The cities on the map are at sea level, the place where land is the same height as the sea.

   a. List the cities on the map that are at sea level.

   Seatle, Washington San Fransico, Los Angeles, San Diego, California   Honolulu, Hawaii
   Tokyo, Japan

   b. Locate and circle Japan and Hawaii.

2. The colors on the color-code strip show the depths of the ocean.

   a. Circle the color on the strip that represents 20,000 feet or more.

   b. Draw an arrow from the strip you circled to cities on the map that are near that color.

3. Based on the fact that tsunamis originate in the deepest parts of the Pacific Ocean, in what places are earthquakes and volcanic eruptions likely to occur?

   Tokoyo
   Japan and   Aleutian Islands   Honolulu

B. More than three fourths of the world's earthquakes occur around the Pacific Ocean. This area around the Pacific Ocean is called the "Ring of Fire" because of its frequent volcanic eruptions.

1. Describe how tsunamis, earthquakes, and volcanic eruptions are connected.

   A volcanic eruption can send ashes into the ocean causing an under water earthquake creating a tsunamis.

2. Explain why Japan, Alaska, and Hawaii's relative location makes them likely targets for tsunamis.

   The ocean water near them is very deep where most tsunamis originage, because of the deep water.

Lesson 2

ACTIVITY
Create a model of a tsunami and compare its effects with those of a hurricane or typhoon.

Creating a Tsunami

The **BIG** Geographic Question

How do the effects of a tsunami compare with those of a hurricane or typhoon?

From the article you learned what a tsunami is and what causes it. The map skills lesson showed you how to identify ocean depths, which affect tsunamis. Now create a model of a tsunami and do research to show how it compares to a hurricane or a typhoon.

A. Like a tsunami, hurricanes and typhoons are also natural hazards that form in the oceans. Look in the Almanac to find out more about a hurricane or a typhoon and answer the following questions.

1. How does a hurricane or typhoon form? _____

_____

_____

2. How big can it get? _____

_____

_____

3. How does water move as the hurricane or typhoon occurs? _____

_____

_____

4. What effects can a hurricane or typhoon have? _____

_____

_____

_____

B. Compare what you have found out about hurricanes or typhoons with what you know about tsunamis, and complete the chart below.

Natural Hazard	Tsunami	Hurricane/Typhoon
Cause		
Size		
Effect in Water		
Effect on Land		

C. Assemble the materials you will need to create your three-dimensional model of a tsunami.

1. Look again at the diagram of a tsunami on pages 8–9. Then think about what kinds of materials—cardboard, paper, clay, papier-mâché, styrofoam—would work best to show each stage of a tsunami and its effects.

2. Display your model in a box, on a flat board, or on cardboard. Make labels that explain what a tsunami is and its different stages.

Lesson 3

Onward, March!

**As you read notice the problems
that the physical features of a strange land
presented for Xenophon's army.**

In 401 B.C., about 2,400 years ago, a young Greek named Xenophon signed on as a mercenary, a paid soldier in a foreign army. The army, consisting mostly of Greeks, was organized by a Persian prince named Cyrus. Cyrus planned to capture the Persian Empire from his brother. The huge Persian empire extended across the whole of Asia Minor (present-day Turkey) to present-day India. A soldier could die if he were left to find his way home in this hostile, foreign land. But Xenophon was an adventurer. Ignoring his teacher Socrates' advice, he began the journey east.

When Xenophon and Cyrus' soldiers arrived in Persia, they found the king's army waiting for them near Babylon. The two sides clashed at Cunaxa, but the Persians were no match for the disciplined Greek warriors. The Greeks outfought the Persians, but in the process, Cyrus was killed and so were many of his top officers, leaving the Greeks leaderless. Young Xenophon was chosen to lead the men back to Greece.

The journey would not be easy. One thousand miles of desert, mountains, and rivers separated the men from the Aegean Sea. The terrain and the wildlife were foreign to them. But Xenophon was a skilled soldier. He knew he must use scouts to determine the geography of the regions that lay ahead. It was their job to find passes through the rugged mountains and to help navigate over the plains.

A bust of Xenophon

14

The soldiers' problems were not restricted to the physical geography through which they passed. The people who lived in the high mountains and hidden valleys of the region feared the fierce army. The Kurds, a mountain group from a land near the present borders of Iraq, Iran, and Turkey, were particularly unfriendly. They took their families into the hills and tried to stop the Greeks from moving toward Armenia. When Xenophon's army finally pushed through the mountain pass, they still had to cross the deep River Centrites. The Greeks were able to cross the river after a scout discovered a shallow area.

Xenophon and his army come upon a mountain pass.

Not all of Xenophon's encounters with the local peoples were bad. Occasionally local guides directed the army through unfamiliar terrain. These guides helped them through Armenia, over the mountains, and to the Black Sea. A seafaring people, the Greeks celebrated when they saw the water. They could sail quickly to Byzantium, which is now Istanbul, and make their way home.

When Xenophon returned to Greece, he recounted his adventures in *The Anabasis*, known as *The March of the Ten Thousand.* His story has entertained millions and has even influenced our language. The word *xenophobia*, meaning "a fear of strangers," is derived from Xenophon's name. *Anabasis* is a word used today to refer to a military advance.

Lesson 3
MAP SKILLS Reading Historical Maps

The map shown here traces the route Xenophon traveled when he withdrew across Asia Minor. Study the physical features and relative location of places to get an idea of the journey Xenophon and his men undertook. Use the map scale to help you measure distances.

Xenophon's March Across Asia Minor

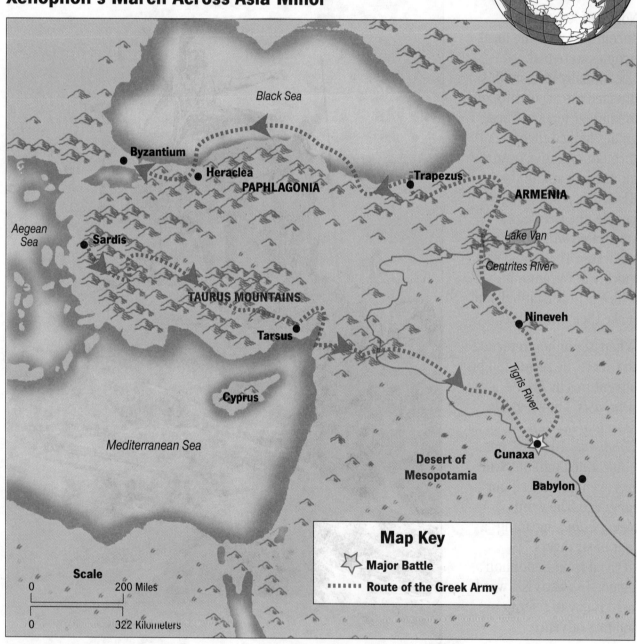

A. **Trace the route of Xenophon from Cunaxa to Byzantium. Use a piece of string to measure on the map the distance between the following places along the route. Then use a ruler to measure the string in inches. Use the map scale to convert inches to miles.**

Distance Between   **Number of Miles**

1. Cunaxa and Nineveh   ~~300~~ 400

2. Nineveh and Trapezous   ~~400~~ 600

3. Trapezous and Byzantium   ~~600~~ 800

B. **Use information from the map to answer the following questions.**

1. What was the closest city to the battle at Cunaxa? Babylon

2. What river did Xenophon's army follow from Cunaxa to Nineveh? Tigris

3. In what desert did the battle of Cunaxa take place? Mesopotamia

4. What was the first mountain chain the army crossed on its journey to Cunaxa? Taurus Mountians

5. What body of water did the Greek army sail to reach Heraclea? Black Sea

C. **Draw conclusions about Xenophon's route by using information from the map.**

1. Note the **topography**, or physical features, along the route. What was probably the easiest part of the journey to travel on foot?

   In the Desert of Mesopotamia

2. Why might the trip north from Nineveh have been particularly difficult?

   Crossing the Tigris River

3. Why would the fact that Asia Minor is located on a peninsula have helped the Greeks in their journey?

   They could travel by sea eaiser because Asia Minor was covered by 3 bodies of water.

17

Lesson 3

ACTIVITY
Explain how a military event in history was affected by xenophobia.

The Significance of Xenophobia

The **BIG** Geographic Question

What physical and human elements influenced the outcome of Xenophon's march?

In the article you read about how Xenophon's army faced many problems as it struggled to get back to Greece from Persia. In the map skills lesson you learned about how the land's physical features made it difficult for the army to travel. Now find out how xenophobia could have affected and changed the outcome of historical military events.

A. Answer the following questions using information from the article and the map skills lesson.

1. Define *xenophobia*.

   A fear of strangers

2. Describe the physical features of the land through which Xenophon's army passed when it encountered the Kurds.

   Mountians, rivers, and vallys

3. What mountain group was xenophobic? What happened when this group saw Xenophon's large army?

   Kurds, they took their family's to the hills and tried to stop the greeks

4. Why would this mountain group fear the approaching army?

   The army was big and powerful

5. How did xenophobia affect the Greeks as they traveled through Persia?

   The needed help from the people that knew the land.

B. Think about the benefits and drawbacks of Xenophon's army and the Kurds fearing each other. Organize your thoughts in the chart below.

Group	Benefits of Xenophobia	Drawbacks of Xenophobia
Xenophon's Army		
Kurds		

C. Write a short skit describing an encounter between the following two historical characters. Answer the following questions to help you plan your skit. Use dialogue to help develop your characters' relationships. Be sure to use historical information from the article.

- a Kurd in Turkey facing troops from Xenophon's army
- a Greek mercenary serving in the army with Xenophon facing a group of Kurds

1. How do you feel about meeting these strangers? Do you fear them? Why or why not?
2. What would you say to them?
3. Would you try to help or hurt these strangers? How could you help them?
4. How would your actions be different from or similar to those taken during Xenophon's time?

D. What do you think was the importance of xenophobia in the historical march of Xenophon?

_____

_____

_____

_____

_____

And The Winner Is...

As you read about the two great cities of the ancient world that wanted to control the markets and rich soils of the western Mediterranean, think about which one would win control and why.

In the fifth and fourth centuries B.C. the city of Carthage dominated the western Mediterranean region. Established by the Phoenicians of present-day Lebanon, Carthage was a rich and powerful city in northern Africa and southwestern Europe.

How had it become so powerful? The Carthaginians were master shipbuilders and skilled sailors. Their ships could sail where others had never gone. Carthaginian fleets explored the Mediterranean, then went through the Strait of Gibraltar, and down the west coast of Africa. Along the way, they established colonies. The colonies in Africa, the islands of the Mediterranean, and Hispania (present-day Spain) helped Carthage expand its growing empire and provided markets for corn, wheat, and other crops its farmers grew.

In the third century B.C. Carthage's power was threatened. The relatively new city of Rome was growing. Eager to establish its own markets and to expand its territory, Rome envied Carthage's colonies and its control of the Strait of Gibraltar.

The cities of Rome and Carthage fought for control of the western Mediterranean region.

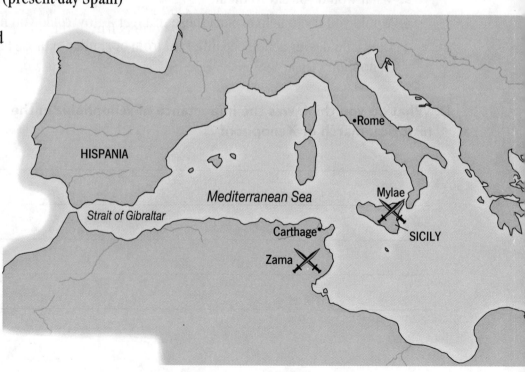

Roman and Carthaginian battleships were called *quinqueremes* because they were powered by five sets of oarsmen on five decks.

The two powers first clashed in 260 B.C. Though the Romans had never had a navy, they studied Carthaginian ships and built warships to equal Carthage's fleet. At the Battle of Mylae in Sicily, the powerful Carthaginian navy was defeated by the newcomers. This was the beginning of what were called the Punic Wars. It was also the beginning of Carthage's decline and Rome's rise to power.

Carthage continued to challenge Rome and to try to protect its monopoly, or complete control, of the western Mediterranean markets. But Rome was growing too strong and was determined to expand its borders. The Second Punic War, which began in 221 B.C., ended with the Battle of Zama in 202 B.C. Zama was located about one hundred miles southwest of Carthage. The Carthaginian army was fighting to protect Carthage, but the Roman army defeated them at Zama, then marched into Carthage. The two enemies signed a treaty that stripped the defeated Carthage of much of its wealth and power.

Not much was left of the once-great Carthage. Nevertheless, Rome wanted to make certain the city never struck again. Romans sought any reason to attack Carthage and found one in 149 B.C. The Third Punic War ended in another Roman victory. This time Rome made sure that Carthage would never again be a threat. The Roman army burned the city, slaughtered its people, and took the few survivors as slaves. To complete the city's destruction, the Romans plowed salt into the soil to make certain nothing ever grew there again. Carthage was dead. A new superpower was born.

Lesson 4
MAP SKILLS Comparing Maps of Different Time Periods

Both of the maps on this page show the same area. The difference between them, however, is about 2,300 years. Maps can show features such as rivers, oceans, and landmasses. They can also be used to indicate human-made places and boundaries, such as cities, states, and countries.

Map 1: Western Mediterranean in the Third Century B.C.

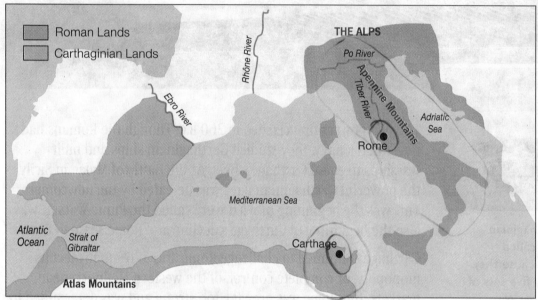

Map 2: Western Mediterranean Today

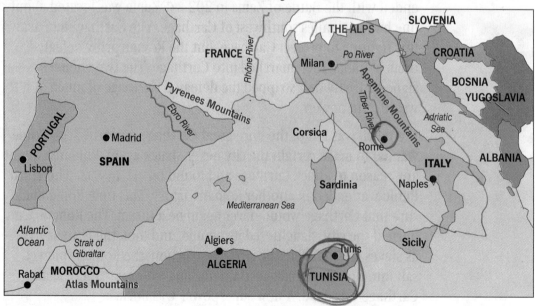

A. **Study the two maps. Circle the following places on both maps.**

  1. Rome
  2. Carthage

  3. Italy
  4. Tunisia

B. **Study both maps to answer the following questions.**

  1. In which modern country is the city of Rome located? _Italy_

  2. Locate the ancient city of Carthage. In which modern country was it located? _Tunisia_

  3. What present-day city appears to be located in the same place as ancient Carthage?

  Tunis

  4. Into which modern European country did Carthage's power extend before the

  Punic Wars? _Spain_

  5. Carthage colonized along the coastlines of which three northern African countries?

  Morocco   _Algeria_   _Tunisia_

  6. What two modern countries does the Strait of Gibraltar separate?

  Morocco   _Spain_

C. **Compare the information on the two maps to draw conclusions about how they differ.**

  1. How does the area of Roman domination compare on the two maps?

  Rome domonated most of Italy, and now it is just one city

  2. How does the area of Carthaginian domination compare on the two maps?

  It covered northern Africa and some of Spain, now it dosn't exist.

  3. What boundaries exist on the modern map that do not exist on the ancient map?

  Europe and African countries.

  4. Given what you learned in the article, why do you think Carthage does not appear on the second map?

  It was destroyed in 149 B.C.

Lesson 4

ACTIVITY Create a resource map of the Carthaginian empire.

When Carthage Was Wealthy

The **BIG** Geographic Question

What resources contributed to the wealth of Carthage?

You learned from the article about the wealth and power of the ancient city of Carthage. The map skills lesson showed the area Carthage ruled. Find out how the city's location and its access to natural resources contributed to its wealth and power.

A. Use information from the article and map skills lesson to answer the following questions.

1. Which modern countries occupy the area that Carthage covered?

   Spain, Morocco, Algeria, Italy, and
   Tunisa

2. What are the major physical features of the area surrounding the ancient city of Carthage?

   The Mediterranean Sea, the Strait of
   Gibraltar, Northern Africa, Southwestern
   Europe.

3. How did these major physical features help Carthage maintain its position of wealth and power?

   Carthage was always near water
   because they were good ship builders
   and sailers.

B. Using the Almanac find out about the resources of the Carthaginian Empire. In the space below draw a symbol to represent each resource.

C. Use the information you've collected on physical features and resources to fill in the map of the Carthaginian Empire in the Almanac. Put the symbols you created in the map key. You may wish to color-code the physical features and provide a key to each color's meaning. Don't forget to give your map an appropriate title.

D. Explain what your map shows about how Carthage became powerful and wealthy.

Lesson 5
Europe's Cities

As you read about how cities develop, think about how their locations influence both their function and characteristics.

The function of a city is influenced by its location, resources, accessibility, and inhabitants. Many of the cities in Europe today were built centuries ago. From 200 B.C. to A.D. 400, much of Europe was part of the vast Roman Empire. After the fall of Rome, other cities began to grow in Europe. There were different reasons for the growth of cities, ranging from defensive needs to the need for transportation links for trade.

London developed as a fortress city on the Thames River.

One important reason for the development of some cities in Europe was the need for fortresses to protect the surrounding area. Cities that were located in an easily defensible position, such as atop a hill or on a river, were needed. A city situated on a hill provided a perfect lookout for outsiders or invaders. Cities located on a river played an important role as links for transportation and trade. A fortress located on a river could easily defend itself against troops attacking from the water because it could sink the invaders' ships.

Some cities began as trading centers at crossroads and grew into market towns. Other cities developed because their location on a body of water attracted merchants who used the waterways to ship and receive goods from all over Europe. Financial institutions developed as a result of the growing number of business exchanges.

As Europe's population grew, the need for local government in different regions became important. Cities that became centers of government were located so that most citizens could travel to them easily. As Europe was organized into countries, many of these centers of government became the capital cities of the countries.

Cities such as **Rome** became religious gathering places.

Religion was another reason for the growth of cities in Europe. Some cities became religious centers after being centers of trade and government. Because many people who came to the cities for business were exposed to religious teachings, churches were able to grow.

Resources such as people and money were also concentrated in the cities. As a result, hospitals and universities developed. Hospitals provided important services to growing cities, as people traveled to receive current medical treatments. Universities were established to accommodate the people who came to the cities to exchange ideas and to learn.

Cities also grew into cultural centers. People gathered in cities and found ways to express themselves through the arts. Diverse populations brought their own music, literature, and other forms of artistic expression to share with one another. Museums, opera houses, art galleries, and libraries provided places where ideas could be shared.

There are many factors that contribute to the growth of cities. They can grow because of defensive needs, business and trade activities, and government. Other factors such as religion, health and educational services, and culture help promote the growth of cities.

Cities such as **Paris** became cultural centers where art and music flourished.

Lesson 5

MAP SKILLS Using a Map to Study City Locations

Maps show where places are located. The political boundaries of countries, states, and cities are usually included on maps. The dots on a map usually designate cities. A star usually designates a capital city.

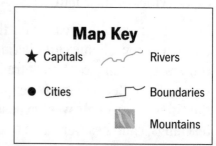

Map Key

★ Capitals ⌇ Rivers

● Cities ⌐ Boundaries

▲ Mountains

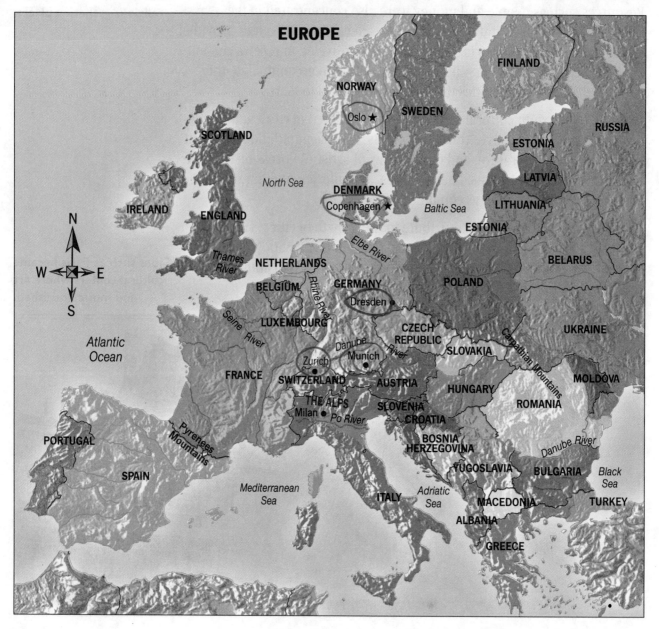

EUROPE

A. Look at the map of Europe. Circle the following cities and study their locations.

√ 1. Zurich, Switzerland   √ 4. Copenhagen, Denmark
√ 2. Munich, Germany   √ 5. Dresden, Germany
√ 3. Oslo, Norway

B. Answer the following questions about the cities listed above.

1. Which of the five cities are landlocked? _Munich, Zurich, Dresden_

2. What large bodies of water are the remaining two cities located near? _____
Baltic Sea and North Sea

3. Between the two cities located near large bodies of water, which would you consider to be the most favorably located? _Oslo_

4. Why is the location of this city a favorable location? _It is in a bay making it easy for people to get in and out._

C. Analyze and indicate each city's location on the chart below. Then indicate on the chart the advantages and disadvantages of cities being located close to or far from these features.

Cities	Location	Advantages	Disadvantages
Zurich, Switzerland	Near Alps and Rhine river. Norther Switz.	Mts. on one side river on other	Near another countries border
Munich, Germany	Near Danube and Foothills of the Alps Southern Germany	In middle of Mts.	▽
Oslo, Norway	On North Sea Southeastern Norway	Easy In and out. Trade from sea	
Copenhagen, Denmark	On Baltic Sea East side of Denmark	Not far from more trade access to water	Attack able by water
Dresden, Germany	On Elbe river Eastern Germany	In Mts. Near river.	

Lesson 5

 **ACTIVITY**   Identify the reasons for the growth of a
city in Europe today.

A City on the Rise

 The **BIG**   **What factors might influence the
growth of a city in Europe today?**
Geographic Question

**From the article you learned why and how cities developed in Europe in
the past. The map skills lesson helped you locate cities of Europe in relation
to nearby physical features. Now determine why and how Milan, Italy, has
become one of the fastest growing cities in Europe today.**

**A. Use what you have learned about how cities developed in the past to answer
the following questions.**

1. Name four characteristics that influence the function of a city.

   _____   _____

   _____   _____

2. Name two reasons why rivers were important to the location of cities.

   _____

   _____

   _____

   _____

3. How did population growth help the development of capital cities in Europe?

   _____

   _____

   _____

   _____

4. Why were hospitals and universities established as European cities grew?

_____

_____

_____

B. Using the information from the article, the map skills, and the Almanac, complete the following chart about Milan.

Physical Features	
Economic Features	
Political Features	
Cultural Features	
Population Features	

C. Use the information you have gathered to write a paragraph about why you think Milan is rapidly growing as a leading European city. Think about its location and how it affects the activities within the city.

_____

_____

_____

_____

Lesson 6

Europe's Rivers

As you read about the rivers of Europe, think about how they affected Europe's growth.

A river can have a significant impact on the development of a region. A glacier, spring, or overflowing lake may be the source of a river. At the other end of the river is its mouth, where the river empties into a larger body of water, such as another river, a lake, or an ocean. All of the smaller channels that carry water into a river, such as rills, brooks, streams, and other rivers, are called **tributaries.** The primary function of a river is to drain a region. A river and its tributaries form a river system. The area drained by a river system is called the system's **watershed,** or drainage basin. This area often contains some of the region's most fertile land.

Because of Europe's major rivers and extensive coastline, water has long been an important means of transportation. The large number of rivers has provided easy transport for people and goods into and out of Europe's interior, whether for reasons of commerce, conquest, or colonization.

The Rhine and Rhône rivers are located in western Europe. The Rhine is an important European river because it is easy to navigate and is widely used by the large population that lives within its

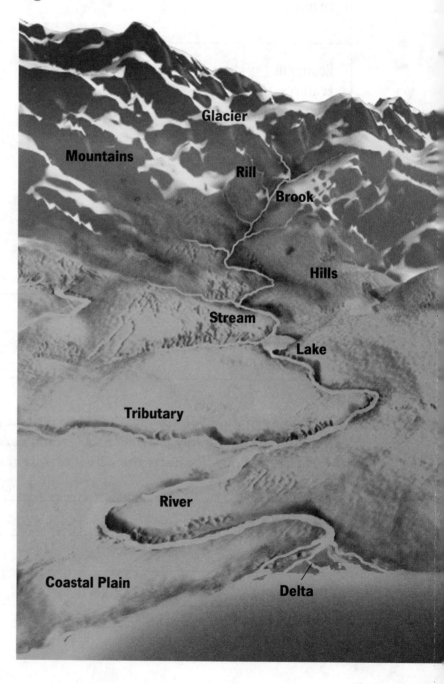

Glacier

Mountains

Rill

Brook

Hills

Stream

Lake

Tributary

River

Coastal Plain

Delta

The Rhine is one of Europe's most important rivers.

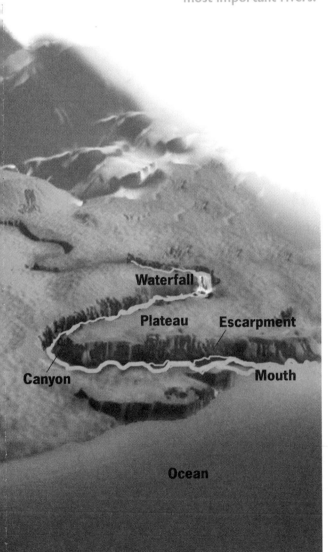

basin. The river begins in the Alpine Mountains (Alps) and is fed by melting snow. As it flows, it forms part of the borders of Switzerland, Liechtenstein, Austria, France, and Germany. It then flows through Germany and the Netherlands into the North Sea. The Rhine receives water from several smaller rivers, such as the Main, Lippe, and Moselle. Along its banks are several major cities, including Cologne, Bonn, Strasbourg, and Rotterdam.

The Rhône River also begins in the Alps and flows through Switzerland and France. It is the only major European river that flows directly to the Mediterranean Sea. The Rhône's course cuts deep valleys through the Jura Mountains, located in eastern France. As a result, the river flows in a difficult, zigzag course.

Europe has a greater number of seas and rivers farther inland than any other continent. In contrast, other continents, such as Asia and Africa, do not have the advantage of so many rivers that are easily accessible to shipping. Asia has a system of varied slopes and mountain ranges. This makes the rivers flow in numerous directions, with some flowing north to the Arctic Ocean. Such rivers in Asia are of little use as trade routes because the rivers and part of the Arctic Ocean are frozen much of the year. Another reason for the small number of inland rivers and seas is the size and shape of the Asian continent. It covers a large area of land, has few peninsulas, and is bordered by Europe to the west.

Like Asia, the African continent has few rivers that flow from within its interior. Africa can be described as a landmass of plateaus and **escarpments,** or steep slopes that mark the edges of plateaus. The large central plateau forces rivers to break through the steep slopes near the coast. As a result, the rivers in Africa do not flow far inland. This makes it difficult for sea traffic to use the natural waterways to reach the interior of the continent.

The geography of a region plays an important role in the economic and cultural life of a region. Physical features such as river systems can enhance, as in the case of Europe, or hinder, as in the cases of Asia and Africa, the growth and development of a region.

Lesson 6
MAP SKILLS Using a Map to Learn More About Rivers

Maps can show the physical features of areas, such as lakes and rivers. Usually, the course of a river is shown on a physical map. This includes the river's source (where it begins), the river's mouth (where it flows into a larger body of water), and important cities located along the river. Some of the river's major tributaries might also be shown.

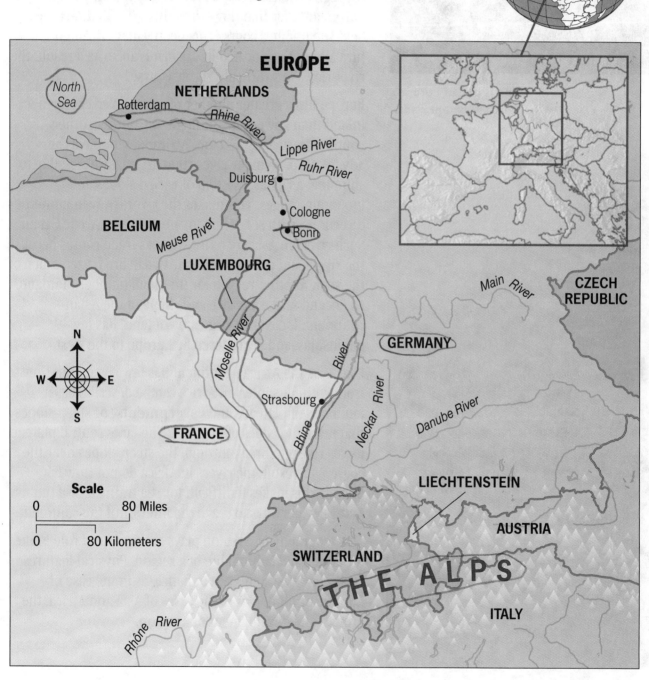

EUROPE

North Sea

NETHERLANDS

Rotterdam

Rhine River

Lippe River

Ruhr River

Duisburg

Cologne

Bonn

BELGIUM

Meuse River

LUXEMBOURG

Moselle River

Main River

CZECH REPUBLIC

GERMANY

Rhine River

Neckar River

Danube River

Strasbourg

FRANCE

Scale

0   80 Miles

0   80 Kilometers

LIECHTENSTEIN

AUSTRIA

SWITZERLAND

THE ALPS

ITALY

Rhône River

A. Look at the map of Europe. Circle the following on the map.

1. Rhine River
2. France
3. Germany
4. The Alps (mountains)
5. North Sea
6. Bonn
7. Moselle River

B. Answer the following questions about the rivers of Europe.

1. What mountain range is the source of the Rhine River? <u>The Alps</u>

2. Which countries does the Rhine River flow through or border? <u>Austria, Netherlands, Liechtenstein, Switzerland, France, Germany</u>

3. What major cities are located on the Rhine River? <u>Strasbourg, Bonn, Cologne, Duisburg, Rotterdam</u>

4. Where is the mouth of the Rhine River? <u>North Sea</u>

5. What smaller rivers flow into the Rhine River? <u>Lippe, Ruhr, Main, Neckar, Moselle, Meuse</u>

6. About how long is the Rhine River? <u>560 miles</u>

7. In what direction does the Rhine River flow? <u>North</u>

8. Where is the Rhine River in relation to the Rhône River? (north, south, east, or west) <u>North</u>

9. Why do you think the Rhine River is so important to Europe? <u>It goes through or borders many countries. It is a long river that goes to the North Sea. It has many major citys on it's banks</u>

35

Lesson 6

ACTIVITY Rate Europe's rivers.

A Rave River Review

The **BIG** Geographic Question

What are the most important features of rivers?

From the article you learned about the rivers of Europe. The map skills lesson showed you the course of the Rhine River and its importance to Europe. Which of Europe's rivers best serves the continent? What characteristics of each river are important?

A. Look at the physical map of Europe and the information about its rivers shown in the Almanac. Then take notes on the following.

Source

1. As is true for most rivers, what is likely the source for Europe's many rivers? _____

_____

Mouth

2. Where are the most useful locations for the mouths of Europe's rivers? _____

_____

Trade

3. What is the best route for a European river for it to be valuable for trade? _____

_____

_____

Population

4. What is the best route for a European river to follow for it to serve the

most people? _____

_____

Tributary

5. What are the best routes for a European river's tributaries for them to have

an important effect on the area through which they flow? _____

_____

B. After reviewing the information about Europe's rivers, complete the chart below to determine the positive and negative features of each river.

River	Positive Features	Negative Features
Volga		
Danube		
Dnieper		
Rhine		
Elbe		
Rhône		

C. Now use the information you have gathered to "grade" each river and determine which is the most valuable to Europe. Compare your "grade" for each river with a classmate's. Did your classmate choose the same or a different river as Europe's most valuable river? Are your "grades" different? Discuss the reasons for the similarities and differences.

1. Volga _____
2. Danube _____
3. Dnieper _____
4. Rhine _____
5. Elbe _____
6. Rhône _____

Lesson 7

Two for One

As you read about Europe and Asia, think about whether their physical features qualify them as two distinct continents.

Unlike other continents, Europe and Asia are not two separate masses of land, but are two parts of the world's largest landmass. In the sixteenth century European mapmakers established them as separate continents. Europe lies on the western portion, extending from the Arctic Ocean to the Mediterranean Sea and from the Atlantic Ocean to the Ural Mountains. Asia extends from the Urals to the Pacific Ocean in the east and from the Arctic Ocean to the tropics near the equator. The Caspian Sea lies between Europe and Asia in a deep depression that dips to 92 feet below sea level in some places. However, no body of water completely separates the two continents, so some geographers consider them one continent, which they call Eurasia.

Completely surrounded by land, the Caspian Sea is really a lake. It lies partly in Europe and partly in Asia—or in Eurasia.

Europe is actually a giant peninsula jutting out from Asia. The land along its coastline curves in and out in a series of smaller peninsulas. Both eastern and western Europe are characterized by a variety of landforms, including mountain ranges, plateaus, and plains. Some of the most spectacular peaks in Europe are found in the Alpine Mountains, which run across southern Europe from the Iberian Peninsula to the Caspian Sea. The North European Plain is a region of flat and rolling land that stretches from the Atlantic Ocean to the Ural Mountains.

Many of Europe's physical features continue through Russia and into China. For example, the North European Plain becomes the West Siberian Plain as it continues east of the Ural Mountains. This plain becomes a plateau region as it extends into Russia and China. The plateaus give way to hills and mountains in Russia, desert regions in China's interior, the North China Plain along the eastern coast, and uplands areas in the southern regions of China.

Many of Europe's climate patterns also extend into Asia. Scandinavia's polar and subarctic conditions continue into the West Siberian Plain. The warm summers and cold winters common across the North European Plain reach into the part of western Asia that lies south of the West Siberian Plain. Climate patterns in eastern Asia, which includes Siberia in the north, are characterized by dry winters and cool, short summers. To the south the climate of China is a mixture of dry, cold conditions in the interior; wet, humid conditions along the coast; and dry, humid conditions in the south.

Asia is the largest of all the continents and makes up about one third of Earth's land surface. Unlike Europe, it is not a peninsula, but a large piece of land surrounded by several groups of islands. The Himalayas, the world's highest mountains, are found in Asia. They stretch across south central Asia between Nepal and China. Mount Everest, Earth's highest point, is found in the Himalaya mountain system.

The northern slopes of the Sayan Mountains are a dramatic backdrop for the southern Siberia landscape.

Though Europe and Asia share the same landmass, they have long been established as two separate continents. The cultures, religious and political beliefs, and languages of the people of these two continents clearly distinguish them. Their size and shape are very different, though many of the physical features that begin in Europe extend through parts of Asia. These facts can help us argue the point of whether Europe and Asia are two distinct continents.

The beautiful cassiopea tetragona flower adorns the Siberian Plains.

Lesson 7
MAP SKILLS
Using a Physical Map to Compare Continents

Topography shown on a map indicates the physical features of an area on Earth's surface. The elevation of mountain ranges, plains, plateaus, depressions, and basins can be identified by contour lines, shading, and color layers. In addition, physical maps show the shape of peninsulas, islands, lakes, coastlines, and borders.

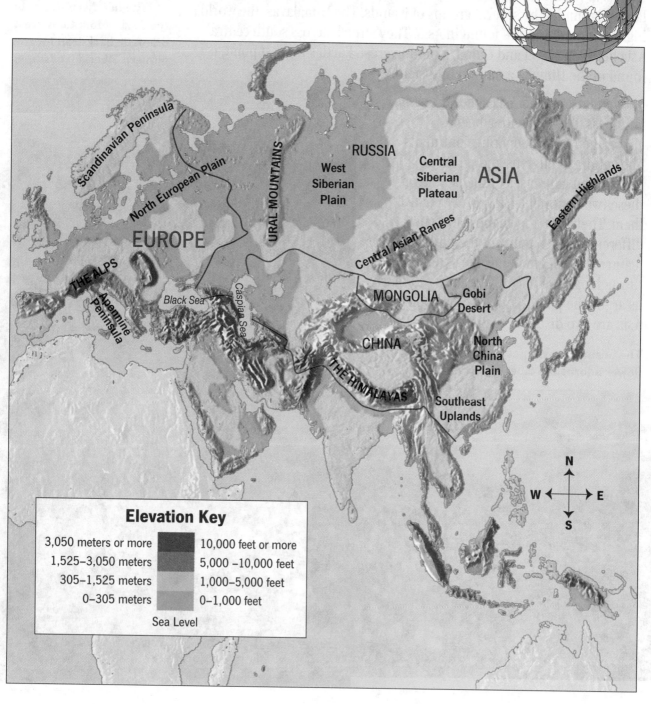

Elevation Key

3,050 meters or more	10,000 feet or more
1,525–3,050 meters	5,000–10,000 feet
305–1,525 meters	1,000–5,000 feet
0–305 meters	0–1,000 feet

Sea Level

A. Study the elevation key on the map of Europe and Asia. Use the key and the map to help you identify the following features.

1. How did you use the map to help you identify:
   a. mountain ranges?

   _____

   b. plains?

   _____

2. Define and distinguish between a peninsula and an island.

   _____

   _____

B. Look at the map and complete the following.

1. Draw a line showing where Europe and Asia are divided.

2. Using the map, write a description of the physical features you see in Europe.

   _____

   _____

   _____

3. Why do you think Europe is considered a peninsula of Asia, rather than Asia being considered a peninsula of Europe?

   _____

4. Write a description of the physical features of Asia.

   _____

   _____

   _____

5. Describe the physical features that are shared by both Europe and Asia.

   _____

   _____

   _____

Lesson 7

ACTIVITY
Organize information to use in a debate, discussion, or report.

Continent Combo?

The **BIG** Geographic Question

Should Europe and Asia be identified as a single continent or two separate continents?

From the article you learned about the diverse physical features of Europe and Asia. In the map skills lesson you looked at some of these physical features on a map. Now decide whether Europe and Asia should be two continents or one.

A. Use the article, the map skills lesson, and your own ideas to answer the following questions.

1. **Geography:** What physical features connect Europe and Asia? What features separate them? Keep in mind the following physical features: plains, mountain ranges, depressions, rivers, and climate zones.

_____

_____

_____

_____

2. **History:** Why have Europe and Asia been considered separate continents?

_____

_____

3. **Culture:** What complications might result if Europe and Asia became one continent?

_____

_____

_____

B. Pretend you are a member of a **World Mapmakers Association.** You are traveling as a delegate to the United Nations to attend a special meeting of their General Assembly. The delegates will present arguments for and against the following proposal. Use the chart below to organize the pros and cons of the proposal.

The United Nations should recognize Europe and Asia as a single continent, hereafter to be identified on all maps as Eurasia.

	Geography	History	Culture
Pros			
Cons			

C. Choose a side on the proposal and organize your arguments in a debate, a discussion, or a short research report.

_____

_____

_____

_____

_____

_____

_____

Lesson 8
The Pearl of Siberia

As you read the story of Lake Baikal, keep in mind how human activities can affect the physical environment.

All Russian children were once taught that Baikal is special. It is the oldest lake on Earth, as well as the deepest. It is the largest freshwater lake on Earth, holding 20 percent of the world's fresh water—more than all of North America's Great Lakes combined. In school children traced the lake's elegant shape and learned the nickname Russians have called it for generations—"the Pearl of Siberia."

Lake Baikal is a natural wonder located in southeastern Siberia, west of the Yablonovy Mountains in Asia. Many species of wildlife can be found only in this lake and the regions surrounding it. Sadly, in the 1960s, due to intensive industrial and resource development, the Soviet Union became largely responsible for the lake's massive pollution. A significant part of Lake Baikal and the surrounding region has suffered from ecosystem collapse as a result of overdevelopment.

All of the living and nonliving things around Lake Baikal form a closed **ecosystem.** This is because all of the lake's water comes from surrounding mountains, and its drainage area is only twice as large as the lake itself. This, along with overdevelopment, upset the delicate balance of the lake's **ecology,** or relationship between the living and nonliving things in the lake environment.

Baikal's decline began in 1896 with the arrival of the Trans-Siberian Railway. Primitive timbering and agricultural techniques followed, causing **erosion,** or the wearing away of soil and rocks by water and wind, on the shore of the lake. A more serious threat came in 1966 when the Baikalsk Pulp and Paper Combine began operation, exploiting the lake's water and the surrounding timberlands.

The clearing of large timber areas devastated much of the surrounding **taiga,** the evergreen forest of Siberia. This also caused erosion and built up **silt,** or fine mud, deposits at Baikal's bottom. Toxic, or poisonous, chemicals filled the lake. Factory waste and untreated sewage, dumped into Baikal from a river flowing into the lake, were made worse by chemical pollutants from another factory downriver. The chemicals from these factories caused serious harm to the human population and the environment.

Once the "crown jewel" of Russia, Baikal has become an environmental battleground. Toxic fallout from air pollutants has damaged about 770 square miles of taiga. **Acid rain,** formed by precipitation and the fumes from burning sulfur and nitrogen oxides, poured onto Lake Baikal. Twenty-three square miles of the lake's bottom were ruined. Pollution has spread so far north that tourist hotels can no longer serve fresh water from the lake. Serious declines have occurred in the size and population of the omul, a species of whitefish that is Baikal's main commercial fish. Continued contamination of the lake and its surrounding area could make the damage to its environment irreversible.

Restoration of this natural treasure could cost the Russian Federation billions of rubles, or Russian dollars, and no one knows where the money will come from. Recently there have been discussions about an international environmental group investing in the lake's future. Factories that pollute Lake Baikal employ thousands of people. If these factories close, workers may have little chance of finding jobs elsewhere. Preventive measures, such as forbidding lumbering close to the lake's shore and monitoring sewage disposal, may give Lake Baikal a chance to heal itself. The life within the depths of the deepest and oldest lake on Earth could indeed have a new beginning.

The nerpa, the world's only freshwater seal, is found nowhere else on Earth except Lake Baikal.

Lesson 8

MAP SKILLS Using Maps to Compare Lakes

The maps shown here can help you compare data about two of the world's largest bodies of fresh water. These maps provide a way to compare Lake Superior in the United States to Lake Baikal in Russia.

Map 1: Lake Superior

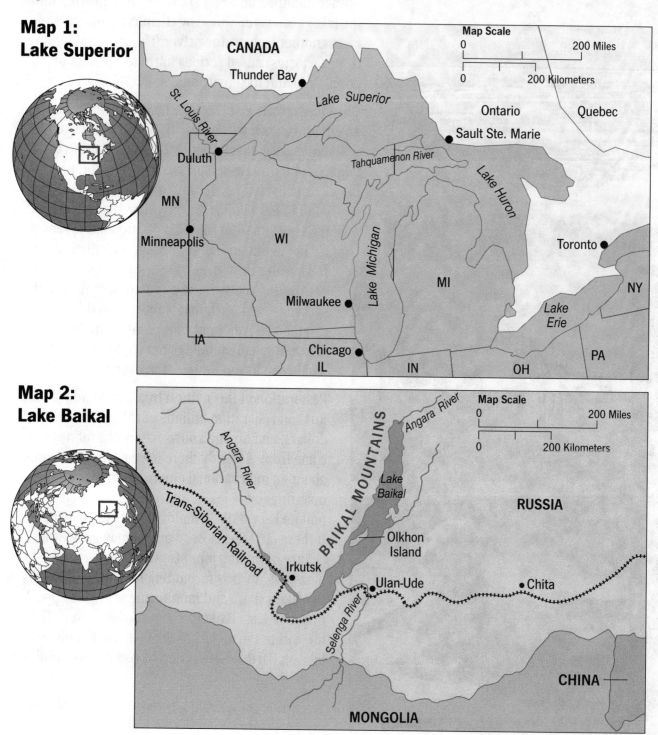

Map 2: Lake Baikal

A. Study the maps and scales to answer the following questions.

1. How long is Lake Superior? _____

2. How long is Lake Baikal? _____

B. Using the maps and and the table below to complete the following.

	Lake Baikal	Lake Superior
Area	12,162 sq. mi.	31,700 sq. mi.
Volume	5,581 cubic mi.	2,916 cubic mi.
Elevation	1,493 ft. above sea level	600 ft. above sea level
Deepest point	5,315 ft.	1,330 ft.
Percentage of Earth's fresh water	20%	10%
Economic uses	harbors, shipping route for minerals, resort	commercial fish, forest and mineral resources, resort

1. Judging from the area and depth of each, which lake appears to be larger? Which lake is actually larger? Explain your reasoning.

   _____

   _____

2. What is the volume of water for each lake? Which lake do you think has a greater percentage of the world's fresh water? Explain.

   _____

   _____

3. Which lake might serve better as a shipping route? Explain.

   _____

   _____

4. Write summary statements explaining the similarities and differences between the two bodies of water.

   _____

   _____

Lesson 8

ACTIVITY

Identify the cause-and-effect relationships associated with the decline of Lake Baikal.

Lake Effects

The **BIG** Geographic Question

How have human activities caused environmental changes at Lake Baikal?

From the article you read about some of the changes in the physical geography of the Lake Baikal area. In the map skills lesson you compared Lake Baikal to another fresh water lake. Now find out how human activity has turned Lake Baikal into an environmental battleground.

A. Find the following terms in the article on Lake Baikal. Write a definition for each term. Then use the term to write your own sentence about Lake Baikal.

1. ecosystem _____

_____

2. ecology _____

_____

3. erosion _____

_____

4. toxic _____

_____

5. acid rain _____

_____

B. Develop a picture of Lake Baikal by completing the following chart of what it was like before and after the arrival of the Trans-Siberian Railway.

Before	After

C. For each change that occurred after the Trans-Siberian Railway came to Lake Baikal, there was an action that caused it. Below, explore the cause-and-effect relationships and make suggestions for how to prevent the problem from continuing in the future.

Cause	Effect	Possible Solutions

Lesson 9
The World of *Islam*

As you read about Islam, think about the role language plays in cultural geography.

Islam is a major religion in many parts of the world. Those who practice it are known as Muslims. Islam, which means "submission to the will of God," began in about A.D. 610. It was founded by a man named Mohammed. He believed he had been sent to warn and guide his people and call them to worship God, the Creator of the whole universe.

Mohammed lived in the Arabian city of Mecca, where he received verbal messages that he believed came directly from God. These messages were gathered into a book known as the Koran. The Koran teaches the Muslim belief in the absolute unity and power of God.

The Koran is written in the Arabic language. There are two forms of Arabic—spoken and written. The written language is considered pure and very beautiful. The Koran represents the highest ideal of writing. The Koran is not simply a holy book. Muslims believe that it contains God's words about laws and how people should live their everyday lives. Because the Koran is so important, many people want to read it for themselves. To do this, they have to learn written Arabic.

The Arab world made great advances in the science of language. Scholars collected and classified words by listening to native Arabic speakers. People began to use the language for philosophy, poetry, and scientific writings.

Over the next hundred years after the death of Mohammed in A.D. 632, Islam spread throughout northern Africa and southwestern Asia. The Arabic language also spread and became the common language of the legal system. Those conquered by the Arabs, such as Syrians, Persians, and Egyptians, had to learn the language to understand the laws

50

and do business with their conquerors. Many people converted to Islam and learned Arabic in order to read the Koran. Migration and trade introduced Islam to people outside of the Arab world. Islam was spread through the movement of Muslims to Indonesia, southern Africa, and the Western Hemisphere in the 1500s and 1600s.

There are five basic practices or duties of Muslims, called the Five Pillars of Islam. First, Muslims must declare their belief in Allah, or God, and accept Mohammed as the prophet. They must understand what they are saying and say it aloud.

Second, all Muslims must pray five times a day facing the direction of Mecca, the holy city of Islam.

Third, charity is required of Muslims in the form of a 2.5 percent tax each year on all that they own. This tax is used to help the poor. It is called almsgiving.

Fourth, during one month of the year, the holy month known as Ramadan, Muslims cannot eat between sunrise and sunset. Muslims believe that fasting cleanses the spirit.

The fifth pillar is called the *hajj.* Once in their lifetime, provided they can afford it, all Muslims must make a pilgrimage to the holy city of Mecca to visit the Kaaba Stone, the most sacred Islamic relic.

Even today children in Islamic countries go to Koranic schools. Beginning at age three, they learn to write by copying verses from the Koran in Arabic. The teacher reads the verses as the children follow along in their copies. Children are expected to memorize many of the 6,000 verses. Eventually the children learn to both read and write Arabic and are given their own copy of the Koran. Although people in different countries speak different Arabic dialects, and other languages as well, written Arabic has come through centuries with little change because of the spread of Islam and the Koran.

Lesson 9
MAP SKILLS Locating Islamic Countries on a Map

Maps can show physical features, population density, and political boundaries of countries. Maps can also be used to show the distribution of a world religion and a language.

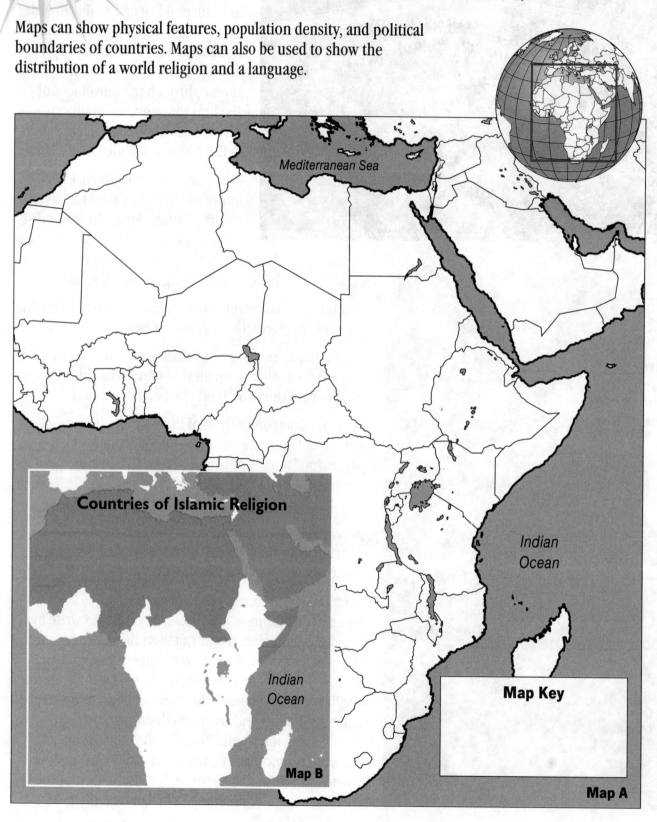

Mediterranean Sea

Countries of Islamic Religion

Indian Ocean

Indian Ocean

Map B

Map Key

Map A

A. Label Map A of northern Africa and the Middle East to show those countries in which Arabic is the main language.

1. Look in the Almanac to find countries in which Arabic is the main language.
2. Using the world map in the Almanac, locate these countries on Map A and label them using a colored marker or pencil.
3. In the map key on Map A, draw and color in a small square using the marker or pencil. Label this square "Arabic-Speaking Countries."

B. Look at Map B to see the areas of northern Africa and the Middle East where Islam is the main religion.

1. Use a different color marker or pencil to outline onto Map A the shaded areas on Map B. Then use the marker or pencil to draw a short line in the map key to represent the outline color. Label this line "Countries of Islamic Religion."
2. Note that you have just indicated on Map A the areas of northern Africa and the Middle East where Islam is the main religion.
3. Note that some countries in the area outlined on Map A are not labeled.
4. Use another different color marker or pencil and the world map in the Almanac to label the remaining countries in the outlined area on Map A.

C. Use your outline map to answer the following questions.

1. Which covers a larger land area, the Islamic religion or the Arabic language?

   _____

2. Which countries included on Map A practice Islam but do not speak Arabic?

   _____

   _____

3. What conclusions can you make about those who live in countries in which Islam is the main religion, but who do not speak the Arabic language?

   _____

   _____

   _____

   _____

Lesson 9

ACTIVITY
Develop research to present information on the spread of Islam.

Islam—A Television Special

The **BIG** Geographic Question

How does a religion spread from one country to another?

In the article you learned about the Arabic language and its role in the growth of the religion of Islam. The map skills lesson helped you see which countries in northern Africa and the Middle East speak Arabic and practice Islam. Now create an informative television special about the Islamic religion.

A. Using the article and information from the Almanac, gather information for a television documentary about the spread of the Islamic religion. As you read, answer these questions.

1. In what city did Islam originate? _____

2. In what country is that city located today? _____

3. Why is that city the most important location in the Islamic world? _____

   _____

4. Who founded Islam? _____

5. What are the five basic practices or duties this founder taught? _____

   _____

   _____

6. In what book can those practices be found? _____

7. When and how did Islam reach areas beyond the Arab world (e.g., Pakistan, Indonesia, China, Malaysia, the United States)?

   _____

   _____

B. Make notes on some of the information you feel would be most important to communicate about Islam.

_____

_____

_____

_____

_____

C. Remember that television is a visual means of communication. Decide what kind of visual aids (map, picture, diagram, and so on) you will use to illustrate each point in the documentary. Note your type of visual aids here.

_____

D. Draw a "storyboard" for your documentary. You can do this by using index cards to sketch one scene at a time. Decide what the viewer will see as each topic is discussed. Write a few lines of dialogue below each sketch. When you have finished the cards, arrange them on a piece of posterboard. Place the cards in the same order that you would present the information in your television special. Use the space below to plan what will on each panel of your storyboard.

Lesson 10

THE STRAIT STORY

As you read about straits, think about their historical and cultural importance to southwest Asia.

Throughout history, **straits**—narrow channels connecting two larger bodies of water—have been of critical concern for defense and trade in southwest Asia. Some of the important straits in southwest Asia and northern Africa are the Bosporus, the Dardanelles, the Strait of Gibraltar, the Strait of Hormuz, and the Suez Canal. (See page 58 for a map that shows the location of all of these straits in relation to each other.)

The Bosporus is a strait connecting the Black Sea and the Sea of Marmara. It separates parts of Asian Turkey from European Turkey. At its narrowest point, it is less than a half mile wide.

Map of Balkan Region, 1993

ROMANIA
YUGOSLAVIA
Black Sea
BULGARIA
Bosporus Strait
MACEDONIA
ISTANBUL
ALBANIA
Dardanelles Strait
Sea of Marmara
GREECE
Aegean Sea
TURKEY
N
Mediterranean Sea

Map Key

🌲🌲🌲 Forest

⛰ Mountains

Prairie

Grassland

The city of Istanbul (formerly Byzantium, and later Constantinople) sits on both sides of the Bosporus. To defend the city from attack by Europeans who wanted to control the Holy Land, fortifications were built along the strait in the fourteenth and fifteenth centuries. Russia has long sought control of this strategic military and trade route from the Black Sea. It provides Russia's only outlet for ships and submarines that is not frozen throughout the year. However, the Montreux Convention of 1936 gave control of the Bosporus to Turkey, which still controls it today.

The Dardanelles lie at the other end of the Sea of Marmara, connecting it with the Aegean Sea. As far back as the twelfth century B.C., the location of this strait has provided an ideal passage between Europe and southwest Asia. In 480 B.C. Xerxes I of Persia built a bridge of boats across the Dardanelles and led an army over it to invade Europe. Later, in 334 B.C., Alexander the Great of Macedonia led his army over a similar bridge to invade Asia.

56

The Strait of Gibraltar, between Spain and Morocco, separates the continents of Europe and Africa. Lying at the western end of the Mediterranean Sea, it is eight miles wide at its narrowest point. The strait is important to world trade because it provides a passage from the Atlantic Ocean to the Mediterranean Sea.

Diagram of the Strait of Gibraltar

The Strait of Hormuz is an important petroleum shipping route linking the Persian Gulf and the Gulf of Oman. It separates Iran on mainland Asia from Oman on the Arabian Peninsula. Throughout recent history, this strait has been a key point in disputes involving oil-producing countries in southwest Asia. The Strait of Hormuz was important during the 1991 Persian Gulf War because it enabled United States troops to enter the Persian Gulf and set up their operations in Saudi Arabia. This allowed troops to be close enough to watch the Iraqi army.

One of the most important waterways in southwest Asia is the Suez Canal, an artificial strait that joins the Mediterranean and Red seas. The canal was cut through 100 miles of desert. It separates the main part of Egypt from the Sinai Peninsula.

Although the Suez Canal is not really a strait, it serves the same function as a strait. The Suez Canal is really a narrow strip of land through which a waterway has been cut. When the Suez Canal opened in 1869, it shortened the distance from Europe to the Indian Ocean and Asia by more than 5,000 miles (8,000 km). During the Six-Day War between Egypt and Israel in 1967, Egypt sank ships to block the waterway, closing the canal for eight years. This prevented Israel from trading and from getting too close to and taking over Egypt's ports and land.

The straits of southwest Asia and north Africa continue to be critical to the trade and strategic defense of many countries. This provides further evidence of the importance of physical geography and how it affects human geography, such as trade, the economy, and politics.

Aerial photo of Suez Canal, Egypt

Lesson 10

MAP SKILLS Locating Features on a Map

You can use a map to locate and look at important straits. The map below shows the important straits in southwest Asia and northern Africa. By studying the map you can see why the straits are important for trade and defense and how these narrow channels of water can affect relationships between places.

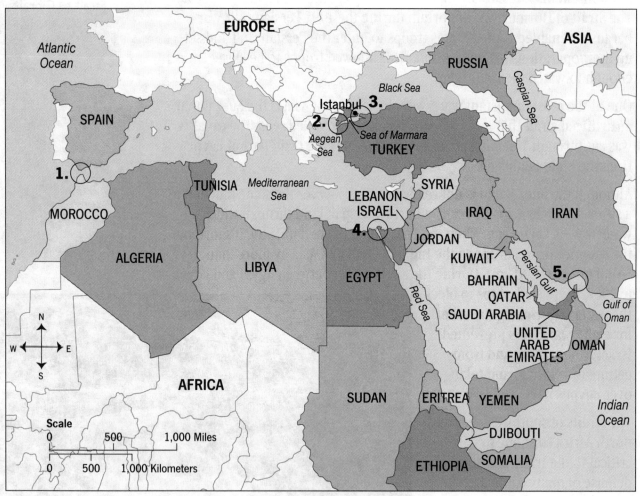

Map Index

1. Strait of Gibraltar
2. Dardanelles Strait
3. Bosporous Strait
4. Suez Canal
5. Strait of Hormuz

A. On the map locate the waterways listed below. Draw a symbol to represent a strait next to each one.

1. Bosporus Strait
2. Dardanelles Strait
3. Strait of Gibraltar
4. Strait of Hormuz
5. Suez Canal

B. Use the article, the map on page 58, and the Almanac to answer the following questions.

1. Would shipping be easier to block in the Gibraltar or Bosporus strait? Why?

_____

_____

2. What countries rely on the Strait of Hormuz to ship their oil? _____

_____

_____

3. Why would Russia want control of the Bosporus Strait? _____

_____

_____

4. How is the Suez Canal different from the straits in this area of the world? _____

_____

5. Why is the Suez Canal important to Israel? _____

_____

6. How was the Strait of Hormuz important to the countries involved in the

   Persian Gulf War? _____

_____

_____

_____

7. Why has the Dardanelles Strait been important throughout history? _____

_____

_____ _____ _____

_____

Lesson 10

ACTIVITY Describe in detail the region's straits.

A Strait Visit

The BIG
Geographic Question

In what ways have straits been important to surrounding countries?

From the article you learned why straits and a canal that serves the same function as a strait are important. The map skills lesson helped you locate several of these waterways and decide how they affect countries around them. Now conduct an in-depth study of one strait.

A. Select one of the major straits in northern Africa or southwest Asia. Use the article, the maps skills lesson, and the Almanac to collect information about that strait. Concentrate and take notes on the following points.

1. Where is the strait located? What bodies of water does it connect?

_____

_____

2. What role has the strait played in the history of the region?

_____

_____

3. Why is the strait important to the politics, economy, and defense of the area?

_____

_____

4. What route would you take to visit the strait?

_____

_____

60

B. Imagine you are writing an entry about your chosen strait for a tourist guidebook. Sketch a map of your strait. Then use the space below to write a description of it and a good route to travel there. Try to make your description informative and interesting.

Description:

_____

_____

_____

_____

_____

_____

_____

_____

_____

Lesson 11

Desert Diary

As you read about how deserts form and change, think about how their geography affects life there.

Nearly 30 percent of Earth is covered with desert. A **desert** is a place where more water is lost to evaporation than falls as precipitation. Many deserts receive less than eight inches of rain per year. Very dry, or **hyperarid,** regions receive less than one inch of rain a year. Many things cause a desert to form. They include wind currents, sources of water, location, topographic features, and human activity.

Most deserts lie in two parallel belts north and south of the equator. Because the sun's energy is strongest at the equator, the air there is heated and rises. As it rises, it cools, releasing moisture as rainfall near the equator. The cooler air, stripped of its moisture, moves toward the poles and sinks toward Earth. This dry, high-pressure air mass blocks moist air from entering. With little rain the land becomes arid, and deserts are born.

The hyperarid **Namib Desert** is a narrow strip along the southwest coast of **Africa**. It is up to **100 miles** wide and more than **1,000 miles** long.

Some deserts go several years without rain. Then one or two large storms will bring all the rain for the year. Flooding can occur, causing erosion and a rapid reshaping of the desert surface. The water cuts channels called **wadis.** Because there are few plants to hold it together, soil, along with other debris, is washed away and deposited in low spots.

Wind is a great shaper of deserts. Huge dune fields, called ***ergs*** in Arabic, are formed by various wind patterns. Three major types of dunes are **barchans,** or horseshoe-shaped dunes, **linear dunes,** and **star dunes.**

About 13 percent of the world's people live in or near deserts. Their activities have, in some cases, increased the size of deserts. Some desert peoples farm its edges. After a few years of cultivation, the soil loses its nutrients. The crops and native plants that have held the soil together and prevented evaporation of moisture from the soil won't grow. The people move their farms to another area, and the cycle repeats.

Sometimes cattle are left to graze until no plants remain. People often cut the few available trees for cooking fuel or clear trees to create more farming land. With no plants left to hold the soil together or retain moisture, new desert land forms. This process is called **desertification.**

Yet another problem is lack of water. More people are digging deep wells to get to water trapped beneath the desert. Some of this water has been there since before the deserts formed. It is called **fossil water.** Once it is gone, it can't be replaced.

Deserts are created by processes of physical and human geography. New desert land may be formed when climate patterns change or as a result of human activities.

Lesson 11
MAP SKILLS
Using a Map to Help Plan
a Trip Route

The map shown below can help you figure out a
journey route that goes through the desert.

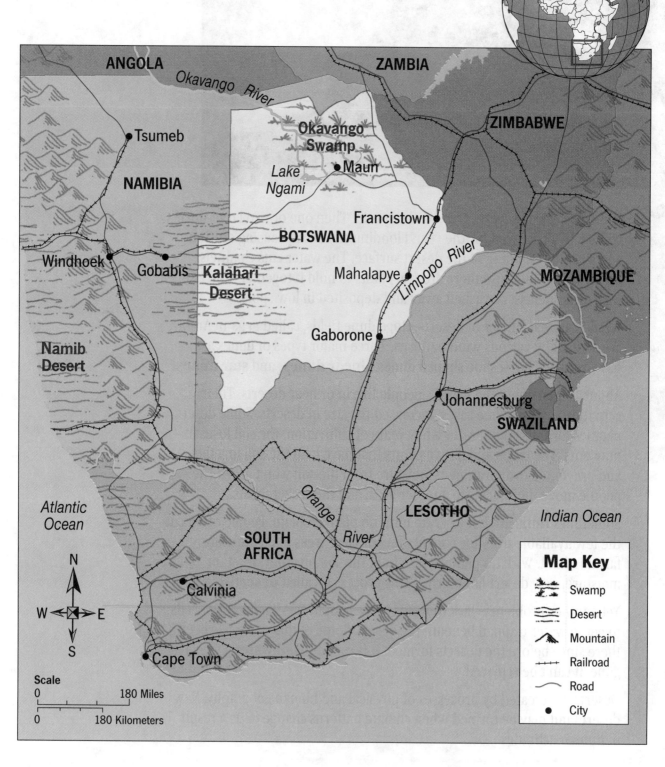

ANGOLA

Okavango River

ZAMBIA

ZIMBABWE

• Tsumeb

Okavango
Swamp

NAMIBIA

*Lake
Ngami*

• Maun

Francistown

BOTSWANA

Limpopo River

MOZAMBIQUE

Windhoek

Gobabis

Kalahari
Desert

Mahalapye

Namib
Desert

Gaborone

Johannesburg

SWAZILAND

*Atlantic
Ocean*

Orange

LESOTHO

Indian Ocean

N

River

W   E

SOUTH
AFRICA

Map Key

• Calvinia

Swamp

Desert

S

Mountain

Railroad

• Cape Town

Road

Scale

0   180 Miles

• City

0   180 Kilometers

A. Study the map. Using different colored pencils for each item listed below, circle the following on the map.

1. Windhoek, Namibia and Gabarone, Botswana
2. Kalahari Desert
3. Towns or settlements along the northern route between Windhoek and Gabarone

B. Use the map, scale, and legend to complete the following.

1. If you were flying from Windhoek directly to Gabarone, how many miles would you travel?

   _____

2. What other two methods of transportation could be used to get from Windhoek to Gabarone? In different colors, trace these two routes on your map.

   _____

3. What form of transportation and route would be the quickest and easiest to travel from Windhoek to Gabarone and how long many miles would the route be?

   _____

   _____

   _____

4. Describe the physical features you would encounter as you travel from east to west.

   _____

5. What supplies would you need to take with you on your trip using the quickest and easiest route? Explain your answer.

   _____

   _____

   _____

6. What form of transportation and route would be the longest and hardest to to travel from Windhoek to Gabarone and how many miles would the route be?

   _____

   _____

   _____

65

Lesson 11

ACTIVITY Show physical and human characteristics of deserts.

Picturing the Desert

The BIG Geographic Question

How are deserts formed, and how do they change over time?

From the article you learned how deserts form and change. The map skills lesson helped you look at a route traveling through Africa's Kalahari Desert. Now make a visual display of a desert.

A. Use information in the Almanac to make two bar graphs showing the precipitation in the Kalahari and Sahara deserts in Africa over the year by completing the following steps.

1. Place the months of the year on the *x*-axis.
2. Place the number of inches of precipitation on the *y*-axis.
3. Appropriately label the axes "Months" and "Inches of Precipitation."
4. Give each graph a title.

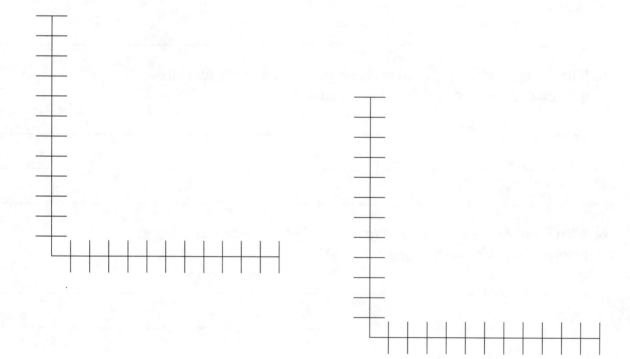

B. Compare the bar graphs you have created by answering the following questions.

   1. Which desert has more rainfall per year? About how much more?

_____

_____

   2. What are some of the natural causes that make the Kalahari a desert? How are these different from the natural causes that make the Sahara a desert region?

_____

_____

C. Using information in the article and the Almanac, select one of the following ideas and create a poster.

_____ **1.** Prepare a poster that shows how nature creates deserts. (Show the effects of physical location, drought, and destruction of plant life.)

_____ **2.** Prepare a poster that shows how people create deserts. (Show how poor farming practices, overgrazing of cattle, and the cutting of trees contribute to decreasing ground cover and increasing desert.)

D. Each poster should have diagrams, pictures, and a few words to explain the main ideas. Plan your poster by taking notes on the following features.

Diagram Ideas	Picture Ideas	Brief Written Description

Lesson 12

A Fishy Story

As you read about Lake Victoria, think about how human activities affect the environment.

Lake Victoria is the second largest freshwater lake in the world. This lake, covering more than 25,000 square miles, lies among the shores of Uganda, Tanzania, and Kenya, Africa. For centuries, local fishers have pulled a rich array of fish from its waters. Millions of Africans were dependent on these fish as their primary source of protein.

The fishing industry depended primarily on a group of fish known as cichlids (si´ klədz). More than 400 species of cichlids lived in Lake Victoria. Ninety-five percent of these species lived nowhere else.

In the 1960s colonial administrators in Uganda were looking for a way to increase their trade with other countries. Although they were a favorite meal for local people, cichlids were too small and bony to be filleted and frozen for export. So the administrators "improved" the lake by adding fast-growing Nile perch. These fish can quickly grow to a weight of several hundred pounds. The perch could easily be processed into fillets, frozen, and shipped to other countries that would pay high prices for them.

Nile perch can grow to six feet.

Cichlids, found primarily in Lake Victoria, only grow to three inches.

68

But the perch proved to be too much of a good thing. To understand why, let's see how a food chain works. The diagram below shows a simple food chain. In a food chain, more complex creatures consume plants or smaller animals. But as the animals get larger, so do their appetites!

What did the Nile perch eat to grow so large? Cichlids! Cichlids once made up at least 80 percent of the weight of fish in Lake Victoria; by 1980 they accounted for less than two percent. By 1985 more than 200 species of cichlids had disappeared.

The countries around the lake now export more than 200,000 tons of perch a year. The cost of perch is so high that local people can no longer afford them, and the cichlids that they once relied on for protein have almost disappeared completely. To make matters worse, the perch have almost completely depleted their own food supply. With few cichlids remaining, the perch have begun eating the lake's tiny brine shrimp. And once those are gone, larger adult perch will begin eating younger ones.

Biologists have said that no other human action in the recent past has put so many species at risk of extinction. By altering the ecosystem of Lake Victoria, the food resources and traditional way of life of an entire people have been threatened.

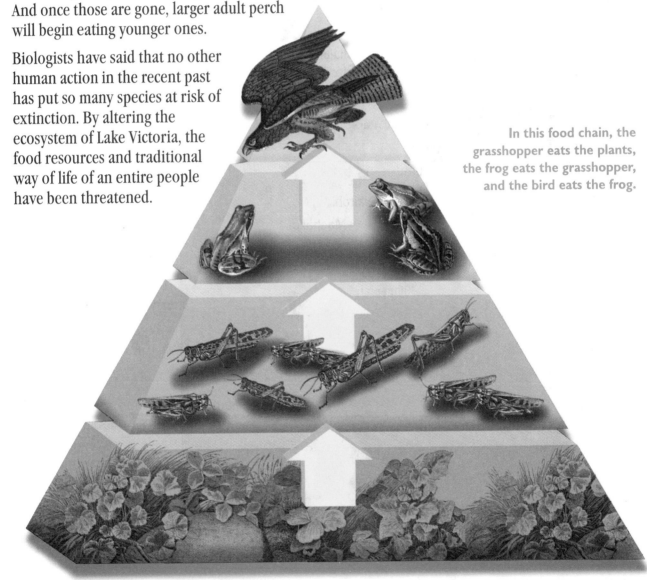

In this food chain, the grasshopper eats the plants, the frog eats the grasshopper, and the bird eats the frog.

Lesson 12
MAP SKILLS
Using a Map to Identify Resources in an Area

The government in Uganda changed the natural balance in Lake Victoria for economic reasons. The decision to add a new species of fish to the lake has resulted in one of the worst possible outcomes. The Nile perch have almost depleted a major food source that many Ugandans and Kenyans have relied on for years. Study the map to decide how other resources might be used to benefit the Lake Victoria people and economy.

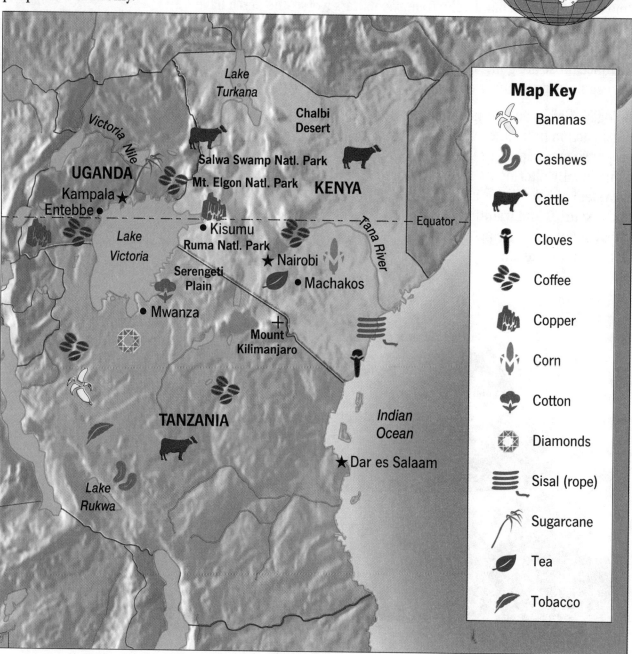

Lake Turkana

Chalbi Desert

Victoria Nile

Salwa Swamp Natl. Park
Mt. Elgon Natl. Park

UGANDA

KENYA

Kampala ★
Entebbe ●

— — — Equator —

Tana River

Lake Victoria

● Kisumu
Ruma Natl. Park

★ Nairobi

Serengeti Plain

● Machakos

● Mwanza

Mount Kilimanjaro

TANZANIA

Indian Ocean

★ Dar es Salaam

Lake Rukwa

Map Key

- Bananas
- Cashews
- Cattle
- Cloves
- Coffee
- Copper
- Corn
- Cotton
- Diamonds
- Sisal (rope)
- Sugarcane
- Tea
- Tobacco

A. Using the map legend, make a list of the resources found in the area of Kenya near Lake Victoria. Remember that resources can include crops, animals, minerals, and natural features that might promote tourism.

Copper

Cattle

Tea

Coffee

corn

Sisal

Cloves

B. Think about the resources found in the Lake Victoria area and answer the following questions.

1. The land used for farming around Lake Victoria is very dry. When there is no rain, crops such as corn will not grow, and the people have no food. What available resource could farmers use to water their corn in times of low rainfall? How would they transport the resource?

erragote their land from the lake water

2. Nile perch were added to Lake Victoria so the area would have a product that could be exported. Describe alternative ways that people might have improved the economy of the area.

Made it a nice lake for tourists. Good fishing, theme parks, nice beaches.

3. List advantages and disadvantages for each of your answers to question 2.

Cost a lot of money to stock a lake, make a theme park. Poulotion would go up. In the long run it would make money and bring in tourists.

Lesson 12

 **ACTIVITY** Develop a representation of a food chain or web.

Creating a Food Chain or Web

The **BIG** Geographic Question

How do the life forms in an area depend upon one another?

In the article you learned how the introduction of Nile perch into Lake Victoria affected the ecology and economy of the area. The maps skills lesson helped you look at other resources of the Lake Victoria area. Now create your own model or representation of the food chain or web of the Lake Victoria area.

A. Before creating your food chain or web, it will be helpful to consider the following questions. Use information from the article and the Almanac to help you answer them.

1. What organisms will be included in your food chain or web of Lake Victoria?

_____

_____

2. How might the removal of one plant or animal from your chain or web affect the feeding relationships?

_____

_____

3. How might the addition of one plant or animal to the chain affect the rest of the chain or web?

_____

_____

4. Suppose environmentalists were writing a book about what has happened at Lake Victoria as a result of the addition of the Nile perch. What title would you suggest for such a book?

_____

_____

5. What lessons can the rest of the world learn from what has happened at Lake Victoria?

_____

_____

_____

6. Research another environmental situation in which a plant or animal was intentionally introduced or unintentionally brought to an area. What kind of effect did it have on the environment? Was it positive or negative?

_____

_____

_____

_____

_____

B. Using art supplies prepare a model of the food chain or web for the Lake Victoria region.

1. Use drawing or construction paper or other art supplies to make three-dimensional representations of the creatures that feed on other animal or plant life. Be sure to include cichlids, Nile perch, insect larvae, brine shrimp, and crocodiles. And don't forget humans! Note that each animal may feed on several of the other smaller species.

2. Arrange the plants and animals that you made on a posterboard. Glue them in place.

3. Label each part of the food chain and give it a title.

C. Imagine one of the elements has been removed from your food chain or web. Below, plan a diagram to show what would happen as a result of its removal.

Lesson 13

Japan
Closes Its Doors

As you read about Japan closing its doors, think about why and how it was able to reopen them to the modern world after centuries of isolation.

Today Japan is a major world power. But at one time, this island nation cut itself off from the rest of the world. For over two centuries, from 1630 to 1853, Japan was closed to all but a few outsiders. How, then, did Japan move from the Middle Ages to the modern world in such a short time? The answer lies in Japan's geography and its people.

In 1603 the Japanese emperor appointed Tokugawa Ieyasu shogun of Japan. *Shogun* means "general," but a shogun's authority was far more than military. Ieyasu took control of the government and united all of Japan under his rule. The emperor lost all power, but he retained his title and his honored position.

Ieyasu was the first of many Tokugawa shoguns. To make sure that people would not revolt, the shogun divided the country into regions, each headed by a daimyo, or feudal lord. The shogun forced the daimyos to take an oath of loyalty. He also passed strict laws to control the movements of daimyos' soldiers, called samurais, and all other people. This prevented people from organizing into groups to overthrow the shogun.

Catholic missionaries came from Europe in the 1500s. Their Japanese converts worried the Tokugawa shoguns. The shoguns feared that the Catholics would obey the rules of their church over the rules of the government. Also, rumors spread that the missionaries were plotting to take over Japan. The Tokugawas ordered all missionaries to leave Japan. They forced all Japanese Christians to give up their faith. Those who did not were persecuted, or put to death.

Organization of Japanese Society

EMPEROR—ceremonial head of government

SHOGUN—ruler of Japan

DAIMYOS

SAMURAI

PEASANTS

ARTISANS

MERCHANTS

From 1630 to 1853 the shoguns closed Japan's doors, continuing to trade only with the Dutch and Chinese. Japan was kept an island unto itself, isolated from the rest of the world. Commodore Matthew Perry of the United States sailed warships with advanced technology to Japan in 1853. The Shogun knew his weapons were not a match for Perry's weapons. He felt forced to sign a trade agreement with the United States. This agreement ended Japan's isolation in 1854 and the Tokugawa shogunate came to an end.

In 1868 the Meiji became the ruling family of Japan. This began a period of rule that moved Japan into the modern world. Meiji leaders were eager to **westernize** Japan by adopting the scientific and technological advances of industrialized countries in Europe and North America. Thousands of Japanese students were sent to the United States, Britain, France, and Germany to study Western military practices, science, government, business, and education. In addition, Western teachers were brought to teach in Japanese schools and universities.

By the end of the Meiji Restoration period in 1912, Japan had built railroads and telegraph lines. It had a constitution, compulsory education, and growing industries. Even after being closed to the world for 200 years, Japan entered the twentieth century as a modern nation ready to build an empire of its own.

Japan's super high-speed bullet train and tranquil tea gardens are good examples of the stark contrast between the Tokugawa and Meiji periods of rule.

Lesson 13
MAP SKILLS
Using Maps to Study How Location
Helped a Country Isolate Itself

You can determine Japan's absolute location on a map by finding its
latitude and longitude. Let's look at how Japan's geography helped make
it possible for the shoguns to isolate the country.

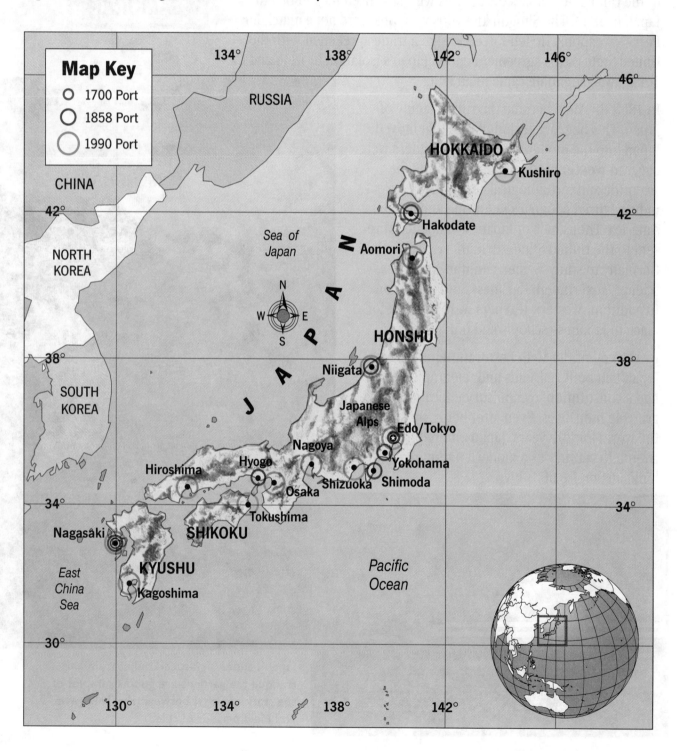

A. Use information from the map of Japan to answer the following questions.

1. What is the current capital of Japan? _____

2. What are its latitude and longitude coordinates? _____

3. What are the four main islands of Japan? _____

_____

4. What bodies of water separate Japan from mainland Asia? _____

_____

5. What body of water separates Japan from North America? _____

6. Where are the Japanese Alps located? _____

_____

B. Use the map to help you answer the following questions.

1. How many ports did Japan have in 1700? Name them. _____

2. How many ports did it have in 1858? Name them. _____

_____

3. How many ports were there in Japan by 1990, and on what side of the

   islands are the majority of them located? Why are they located there? _____

   _____

   _____

   _____

C. Using the information you have learned about Japan, explain how it was able to isolate itself from the rest of the world during the 1600s.

_____

_____

_____

_____

Lesson 13

ACTIVITY

Find out how Japan went from being an isolated country to a modern global power.

Advising the Shogun

The BIG Geographic Question

What roles do physical and cultural geography play in how a country changes?

From the article you learned why and how Japan was able to close its doors from 1630 to 1853. In the map skills lesson you saw how Japan's physical geography made it possible for the shoguns to isolate the country. Now find out how Japan was able to enter the twentieth century as a modern nation.

A. Review the article and construct a time line that chronicles events in Japan from 1603 to 1853.

1603   1630   1853

B. List some physical, economic, and political characteristics of Japan during its period of isolation from 1630 to 1853.

Physical	Economic	Political

C. List some physical, economic, and political characteristics of Japan during the Meiji Restoration Period from 1868 to 1912.

Physical	Economic	Political

D. Compare the attitudes of the Tokugawa Period to those of the Meiji Restoration Period concerning issues of trade, outside ideas, and foreigners.

	Trade	Outside Ideas	Foreigners
Tokugawa Period			
Meiji Restoration Period			

E. Ieyasu, the first shogun of the Tokugawa Period, encouraged trade initially, but later he and other shoguns expelled traders for fear of losing their power. Today Japan is a major trading nation with great economic power. Explain how Ieyasu might respond to present-day Japan.

_____

_____

_____

•

Lesson 14

The Monsoon Season

As you read about monsoons, think about how changes in the physical environment affect human activities.

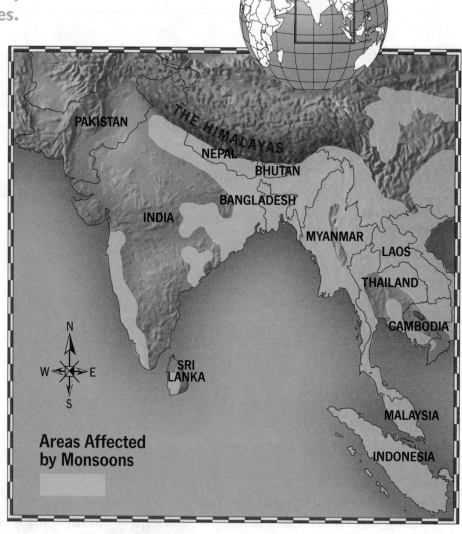

Weather and climate affect the way people live. It influences the kind of clothes they wear, the type of homes they live in, and the food they eat. One type of weather and climate that affects southern Asia is monsoons. A **monsoon** is a wind that changes direction at regular times of the year. The shifting monsoons cause a sharp difference in the amount of rainfall from one season to another.

Monsoons blow over the northern part of the Indian Ocean and surrounding land areas. These winds are a result of the temperature differences in the warming and cooling air over land and water.

From November to March in southern Asia the cool, dry winter air over the inland mountains blows out across the ocean, bringing dry weather. This is a winter monsoon. As the monsoon moves across the ocean, it picks up moisture. During the rest of the year, the monsoon blows from the ocean to the land. It brings warm, moist summer air that results in warm temperatures and lots of rain. The wet summer monsoon is at its height during the months of June through September.

80

The topography of southern Asia affects rainfall levels within the region. Mountains, plateaus, and hills make up most of south Asia. The Himalayas form an almost continuous wall of mountains across the northern boundary of south Asia. The mountain slopes help produce rain by lifting the warm, moist air to a higher altitude. The air cools, forms clouds, and produces rain that falls on the **windward side** of the Himalayas. This is the north side, or the side that faces the direction from which the wind blows. The slopes on the opposite side of the mountain range, the south or **leeward side,** are drier because the winds become dry as they blow over the tops of the mountains. As a result, the southern slopes of the Himalayas get between 200 and 600 inches of rain per year, while the northern slopes average less than ten inches per year.

Rice is one of the world's most important food crops. It requires a warm, rainy climate to grow. Many people of southern Asia are farmers, and many grow rice. Asian farmers grow about 90 percent of the world's rice. The southwest monsoons are very important to the farmers of south Asia. If the monsoons arrive in time and bring enough rain, the farmers' crops will grow, and their families will eat. Sometimes the rains don't come, or they come too late to be beneficial to the farmers. Then there will be little or nothing for the people to eat. If the monsoon causes too much rain, the crops will be ruined and villages destroyed. The summer monsoons can cause rivers to overflow and deposit silt on farmlands. Monsoons are a way of life for people in south Asia. Although the monsoons can cause flooding, the rains they bring are welcomed because they are needed to produce plentiful harvests.

The Himalayas

Lesson 14
MAP SKILLS Using a Map to Link Elevation, Climate, and Vegetation

Maps that show the boundaries of countries and states are called political maps. Political maps can also show locations of mountains, rivers, and lakes.

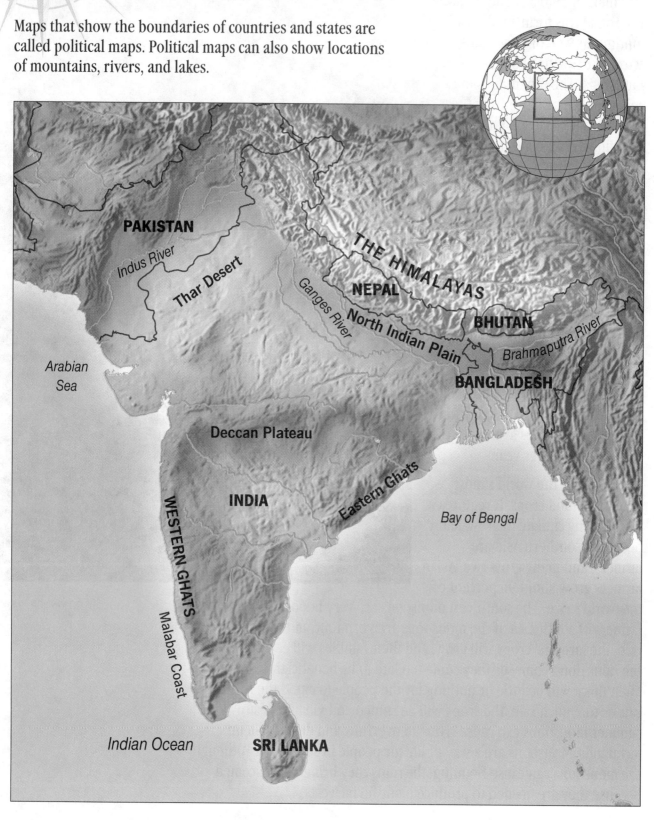

A. Look at the political and physical map of south Asia and complete the following.

1. List the six countries shown.

Pakistan   Nepal

Sri lanka   Bhutan

India   Bangladesh

2. Name the three large bodies of water that border south Asia.

Arabian Sea, Indian Ocean, Bay of Bengal

3. What are the three major mountain ranges that surround India?

Himalyas, Eastern Ghats, Western Ghats

4. What are the three main rivers in south Asia?

Indus, Ganges, Brahmaputra

B. Look for the precipitation and vegetation maps of India in the Almanac. Precipitation maps show the amount of rainfall an area receives. Vegetation maps show how plant life responds to climatic conditions. Complete the following.

1. Which parts of India receive the most rainfall?

2. What type of vegetation is found in the areas with the most rainfall?

3. Describe the general relationship between high levels of rainfall and vegetation.

C. Make connections among what you have learned about rainfall, vegetation, and physical features. Answer the following questions.

1. What kinds of physical features are located on the western coast and north-eastern coasts of India?

2. How do these physical features affect rainfall during the monsoon season?

Lesson 14

ACTIVITY
Learn how the monsoon seasons affect human activity.

Living with Monsoons

How do people adapt to major changes in the weather and climate of their environment?

From the article you learned how monsoons develop and where they occur. In the map skills lesson you learned about the connections among elevation, rainfall, and vegetation and how they relate to monsoons. Now see how the people of the region are affected by these changes in their physical environment.

A. Answer the following questions using information from the article and map skills lesson.

1. Describe the two monsoon seasons. How are they different from each other?

_____

_____

_____

2. Explain how mountains help produce rain.

_____

_____

_____

3. India is a country in south Asia that is greatly affected by monsoons. Why do you think this is so?

_____

_____

_____

84

B. Answer the following questions about how many people in south Asia make their living.

1. On what kind of activity do many people rely to make a living?

   _____

2. What is the main crop produced in south Asia?

   _____

3. How do monsoons affect the crops of farmers in south Asia?

   _____

   _____

   _____

C. Imagine that you are a rice farmer living between the coast and the Western Ghats in southwest India. You have harvested your rice crop, and it is time to take it to market. However, the monsoons are still bringing heavy rains. More than five feet of rain has fallen. You need to sell your rice in order to earn money to buy items that your family needs.

Think about your 30-mile journey to the market. How will the heavy rains affect it? How will you transport your goods? How will you keep your rice dry? What other problems might you encounter?

Write a letter home to your family to let them know about your trip. Describe the journey and how the monsoon affected it.

_____

_____

_____

_____

_____

_____

_____

_____

Lesson 15

A Location Situation

As you read about Singapore, think about how its location has contributed to its economic success.

Singapore is the name of a nation, an island, and a city. The island of Singapore is part of the nation of Singapore, whose capital city is also called Singapore. Located at the southern tip of the Malay Peninsula, Singapore is the smallest nation in southeast Asia, but its size is no indicator of its importance in that region or in the world.

In 1819 the British established a trading post at Singapore. The British were attracted by Singapore's location on the main East-West shipping routes. The **East** refers to regions having a culture derived from ancient non-Europeans, especially Asian areas. The **West** refers to regions having a culture of the Americas or Europe. Singapore's deep natural harbor attracted the British. The Suez Canal also helped Singapore to grow by shortening the length of the East-West routes. Ships no longer had to travel long distances around Africa. The use of steamships to replace sailing ships, which made transportation of goods easier, was another important factor that contributed to Singapore's growth.

The British made Singapore a **free port,** which means that ships passing through the Singapore Strait can unload and reload goods without having to pay import taxes. The advantage of having a free port is that goods can be manufactured as new products, repackaged, or stored with no added taxes.

86

Singapore became the main port for shipping tin and rubber from the Malay Peninsula to the West. However, Singapore wanted to expand its trading role, so it developed its own industries. The country took advantage of its free port status and developed oil refineries, petrochemical plants, and electronics factories. Because of its lack of natural resources, it brought in the raw materials for these industries and shipped the finished products to the surrounding region and to North America and Europe. The textiles, shipbuilding, and ship repairing industries have also grown, and Singapore's banking and financial companies are considered Asia's most stable.

Singapore is a major container port.

Today Singapore has one of the highest standards of living in Asia. Its economy has grown by an average of nine percent every year since 1965, when it became an independent nation. Under a government that not only controls economic growth but actively supports commerce, Singapore has enjoyed long-term political stability, which has encouraged foreign investment in the country. Because of its rapid industrial growth, it is known as one of Asia's "Little Tigers," along with South Korea, Hong Kong, and Taiwan.

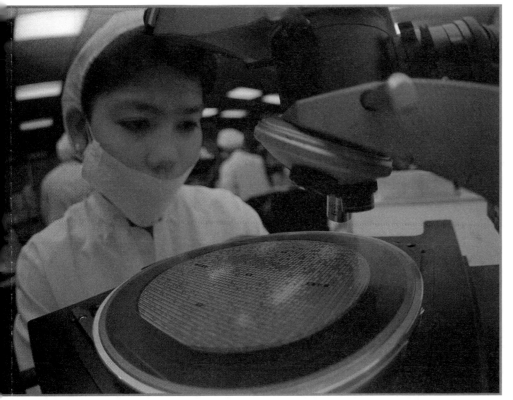

Microchip manufacturing and design in Singapore

Geography has played a major role in the development of Singapore. Having access to East-West waterways, a deep natural harbor, and the ability to develop its own industries has helped Singapore a great deal. It has developed into a strong economic power in southeast Asia and in the world.

Lesson 15

MAP SKILLS
Using a Map to Look at a Country's Location, Land Area, and Management

Many factors influence how a country's economy, government, and interaction with other countries are managed. Whether a country is an island unto itself, a compact mainland peninsula, or a series of sprawling islands are some of those factors.

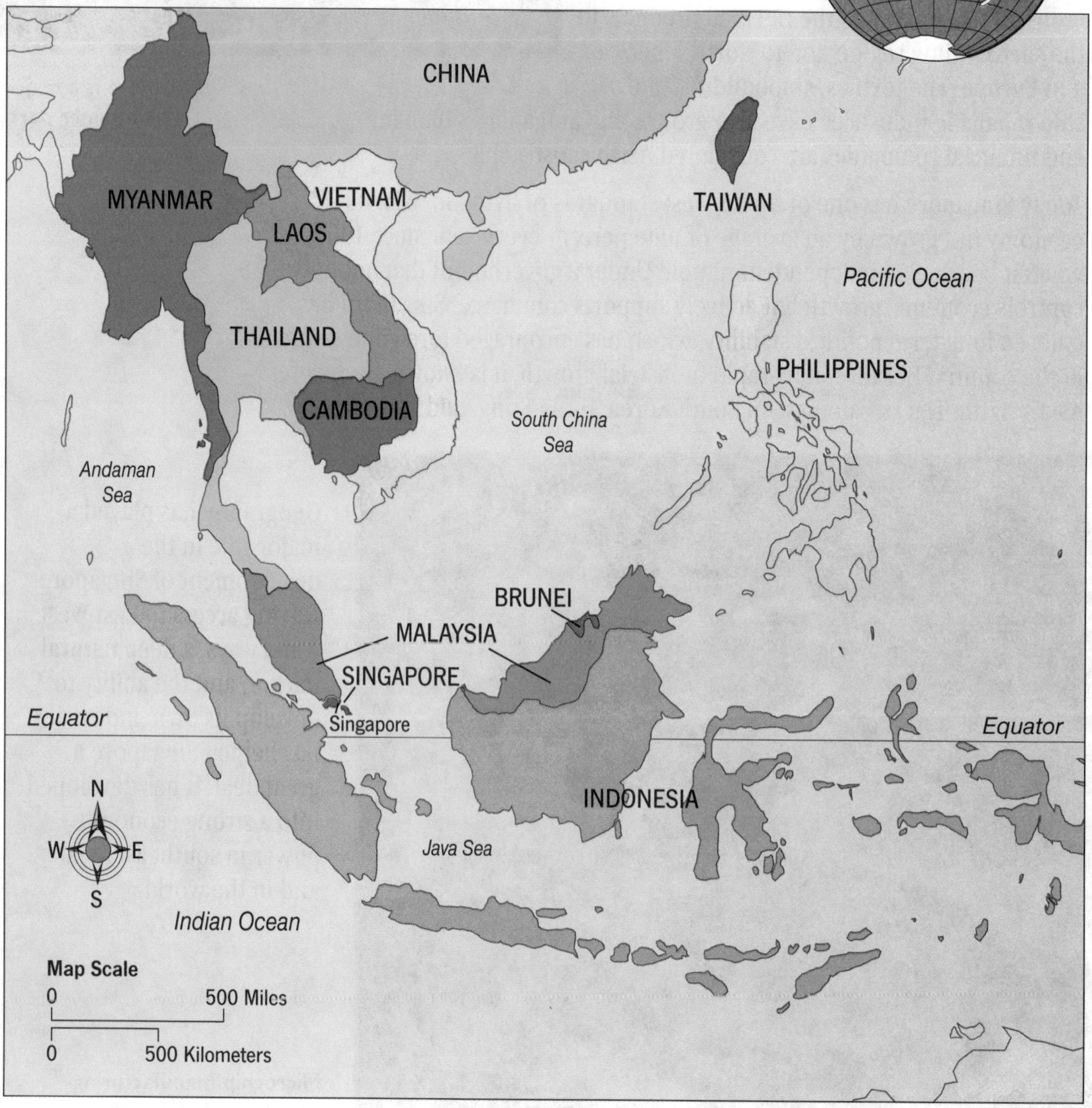

CHINA

MYANMAR

VIETNAM

LAOS

TAIWAN

Pacific Ocean

THAILAND

CAMBODIA

South China
Sea

PHILIPPINES

Andaman
Sea

BRUNEI

MALAYSIA

SINGAPORE

Equator

Singapore

Equator

N

W — E

S

INDONESIA

Java Sea

Indian Ocean

Map Scale

0   500 Miles

0   500 Kilometers

A. Using the article and the map, answer the following questions.

1. What three seas surround Singapore, and to what two oceans does Singapore have access?

_____

_____

2. Why is Singapore important to East-West trade in terms of location?

_____

_____

_____

_____

B. Consider how other countries in the southeast Asia region are affected by their geography by answering the following questions.

1. What countries in the region are small and compact rather than scattered and sprawling?

_____

2. What countries in the region are made up of a series of sprawling islands?

_____

3. What countries in the region are not compact but are relatively large?

_____

C. Use what you learned from the map and your own ideas to answer these questions.

1. Do you think a long, spread-out nation (such as Vietnam) would be harder to unite than a small, compact nation (such as Singapore)? Why or why not?

_____

_____

2. Do you think it would be easier to manage the economy of a small nation (such as Brunei) or a large nation (such as Indonesia)? Why?

_____

_____

Lesson 15
ACTIVITY
Compare Singapore with other major economic powers in the region.

Four "Little Tigers"

The **BIG** Geographic Question

How has geography affected economic power in southeast Asia?

From the article you learned how geography played a major role in the economic development of Singapore. In the map skills lesson you explored whether a country's location and land area affect its management. Now find out how Singapore compares with the other three "Little Tigers"— Hong Kong, South Korea, and Taiwan.

A. The four "Little Tigers" are known for their economic power. Economic power is a country's ability to effectively produce (make), distribute (share), and consume (use) wealth, goods, and services. Use the Almanac to help you complete the chart below and get a picture of the economic power of the four "Little Tigers."

	Singapore	Hong Kong	South Korea	Taiwan
Value of Imports				
Value of Exports				
Major Trading Partners				
Unemployment Rate				
Hourly Wage				

B. Analyze the information on your chart and answer the following questions.

1. Which countries' export values are greater than their import value?

_____

2. Which countries' import and export values almost show a balance?

_____

3. What do the low unemployment rates tell you about each of the "Little Tigers"?

_____

_____

C. Use the world map in the Almanac to identify the location of each of the four "Little Tigers." Then complete the following.

1. Describe the locations of each country.

   a. Singapore: _____

   _____

   b. Hong Kong: _____

   _____

   c. South Korea: _____

   _____

   d. Taiwan: _____

   _____

2. Write a general explanation of how the countries' locations have helped their economic power.

   _____

   _____

D. Imagine you work for the Ministry of Economic Development for one of the "Little Tigers." You have been asked to create a brochure to encourage companies in the United States to invest. Use the article and other reference materials to prepare an outline of the brochure. Remember to use what you know about your country's geography.

Lesson 16

Diversity in the Pacific Islands

As you read about the islands in the Pacific Ocean, think about how they are alike and different.

The Pacific Ocean is the world's greatest natural feature and contains thousands of islands. These scattered islands are called the Pacific Islands, or Oceania. Geographers think the Pacific Ocean has between 20,000 and 30,000 islands. Many are in groups or chains called **archipelagos.** They range in size from the continent of Australia to very small islands. Some cover thousands of square miles; others are tiny piles of rock in the ocean.

The Pacific Islands are divided into four regions: Melanesia, Micronesia, Polynesia, and Australia. A common element to all four regions is that they are surrounded by the biggest and deepest body of water on Earth, the Pacific Ocean. However, the islands have numerous geographic differences.

New Guinea, New Caledonia, Fiji, and Vanuatu are islands in Melanesia. In 1831 French explorer Dumont d'Urville named this area *Melanesia,* meaning "the Black Islands," after seeing their dark, volcanic, jagged forms rising out of the ocean. Most of Melanesia is south of the equator. The Melanesians are related to the aborigines of Australia and are believed to be the first people of Oceania, dating back as far as 3,000 years.

Micronesia means "tiny islands." Guam, the Caroline Islands, the Mariana Islands, and the Marshall Islands are some of the more than 2,000 islands in this group. They lie north of the equator. Micronesians have European or Asian ancestors.

The largest area in the South Pacific is Polynesia. *Polynesia* means "many islands." The Polynesian islands are widely scattered and range from Midway Island on the north to Easter Island on the east and New Zealand in the south. Although New Zealand is part of the Polynesian chain, its culture resembles that of Australia because it was settled by Europeans in the 1700s.

Australia, another area in Oceania, is the only country that is a continent. It is made up of broad plains and plateaus. Its longest and highest mountain range, the Great Dividing Range, runs along the east coast. There are many cattle and sheep ranches and large deposits of coal, gold, and iron ore. On the northeast side of the continent is the Great Barrier Reef, a coral reef that is home to 1,500 species of fish and other marine animals.

Tahiti, aerial view of Bora Bora atoll

There are two main types of islands in the Pacific. The **high islands** are actually the peaks of sunken underwater mountains. The high islands get a lot of rain and are thickly forested. Many of these islands, such as the Hawaiian Islands, have active volcanoes, and earthquakes are a constant threat. New Caledonia, New Britain, the Hawaiian Islands, New Guinea, New Zealand, and Fiji are high islands.

The **low islands** are made of coral reefs that lie on underwater peaks. The reefs are formed from the skeletons of tiny sea animals that have built up over time. Many coral islands are uninhabited because of poor soil or the absence of soil, low rainfall, and lack of fresh water. Only small groups of people can live on some of these islands. These people must depend on fish and coconuts for survival. Most of the low islands are atolls. An **atoll** is a circular or nearly circular coral reef or string of reefs rising above and surrounding a **lagoon,** a shallow area of ocean water.

Almost all of the islands of the Pacific have a warm, tropical climate. Rainfall varies throughout Oceania, but most islands have a wet season and a dry season. An exception to this is Australia. Because of its size and position on the Tropic of Capricorn, Australia's climate is mostly dry due to a globe-circling high-pressure system. As one can see, Oceania has a variety of physical features that make it a unique region on Earth's surface.

Lesson 16

MAP SKILLS

Using a Map to Determine Distance, Direction, and Time

Earth has been divided into 24 time zones. Each time zone forms a belt of longitude (15° across) in which most areas have the same local time. When you cross from one time zone to another, the local time changes by one hour. If you are traveling east, you add one hour for each time zone crossed. If you are traveling west, you subtract one hour for each time zone crossed.

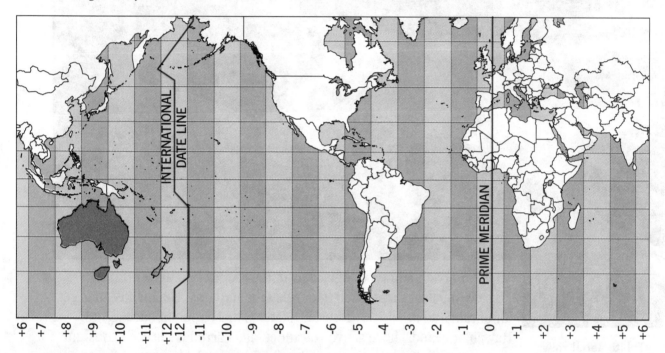

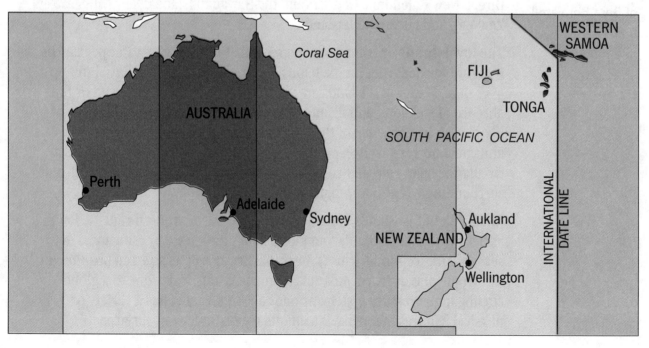

A. Use the time zone map to answer the following questions.

1. How many time zones are there in Australia?

_____

2. If it is 10 P.M. in Sydney, what time is it in Perth?

_____

3. How many time zones are there in New Zealand?

_____

4. If it is 10 P.M. in Sydney, what time is it in Wellington, New Zealand?

_____

The **international date line** is an imaginary line at 180° longitude that runs through the middle of the Pacific Ocean. By international agreement, it marks the spot where each new calendar day begins. Any time you cross the date line going from west to east, you move into the previous day (subtract one day). Anytime you cross the international date line going from east to west, you move into the following day (add one day). The time of day does not change. If you reach the date line at 4 P.M. on Tuesday coming from the east, it will be 4 P.M. Wednesday after you cross the line. If you cross back two hours later, it will be 6 P.M. on Tuesday.

B. With the previous information and the following scenario in mind, answer the questions below.

You are traveling from Western Samoa to Wellington, New Zealand. The trip takes four hours. You begin your trip at 7 A.M. on Saturday.

1. In which direction will you be traveling?

_____

2. How many time zones will you cross during your trip?

_____

3. Is New Zealand east or west of the international date line?

_____

4. What day and time will you arrive in Wellington, New Zealand?

_____

Lesson 16

ACTIVITY
Determine whether the islands of the Pacific form a unified region.

The Sum of Island Parts

> The **BIG** Geographic Question
>
> **What characteristics of an area establishes it as a unified region?**

In the article you learned about the island groups of Oceania. In the map skills lesson you learned where the islands of the Pacific are located in relation to the international date line and how they fall into different time zones. Now determine whether the Pacific Islands form a physically, culturally, and politically unified region.

A region is an area of Earth having one or more common characteristics that are found throughout it. A region can be defined by human factors or by physical features. In each of the four regions of Oceania, the physical geography and history of the people is varied.

A. Choose two countries from one of the four regions of Oceania and compare them. Use the map and Almanac information to complete the chart.

	Country Name _____	Country Name _____
Geographic Area		
Landforms		
Climate		
Crops		
Language		
Population		
Ethnic Groups		
Settlement History		
Government		

96

B. Consider the characteristics of the various types of regions by completing the following questions.

1. What might be some of the characteristics of a physical region?

_____

_____

2. What might be some of the characteristics of a political region?

_____

_____

3. What might be some of the characteristics of a cultural region?

_____

_____

C. Use the information you have collected to answer the following questions.

1. Do islands in the same region all have the same physical characteristics?

_____

_____

_____

2. Do the Pacific Islands form a cultural region? Why or why not?

_____

_____

_____

3. Do the Pacific Islands form a politically united region? Why or why not?

_____

_____

4. Can the area be identified as a physical region? Why or why not?

_____

_____

Lesson 17

Defining Regions

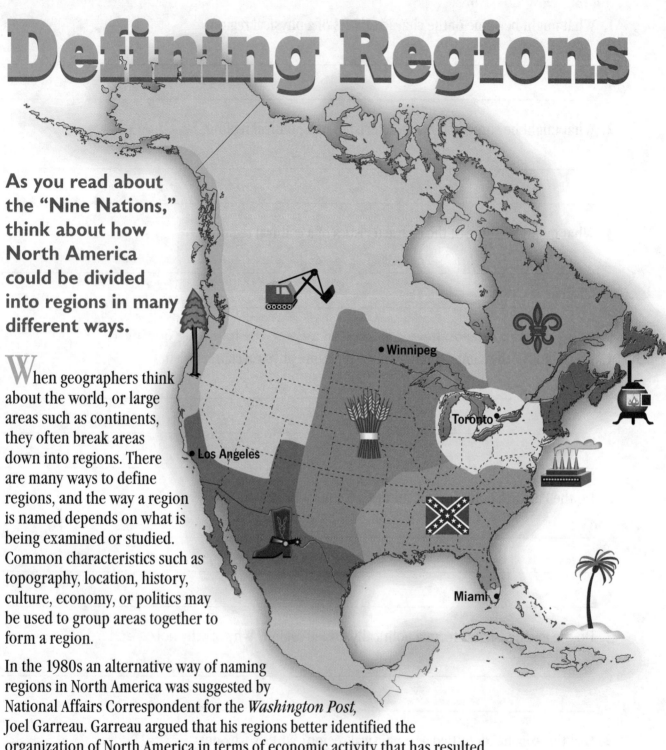

As you read about the "Nine Nations," think about how North America could be divided into regions in many different ways.

When geographers think about the world, or large areas such as continents, they often break areas down into regions. There are many ways to define regions, and the way a region is named depends on what is being examined or studied. Common characteristics such as topography, location, history, culture, economy, or politics may be used to group areas together to form a region.

In the 1980s an alternative way of naming regions in North America was suggested by National Affairs Correspondent for the *Washington Post,* Joel Garreau. Garreau argued that his regions better identified the organization of North America in terms of economic activity that has resulted because of the physical environment. He developed a map of the "Nine Nations of North America." The nine nations concept is valuable because it helps us understand the common characteristics that define each region.

North America can be organized in many ways. Garreau's "Nine Nations" show a unique way of looking at regions.

Breadbasket

The Breadbasket is the vast stretch of land also known as the Great Plains. This region is the heart of agriculture in the United States. It includes the states of Nebraska, Iowa, Kansas, the Dakotas, Minnesota, and large parts of other interior states.

Dixie

Garreau argues that Dixie, the nation located in the southeastern part of the United States, is unified by Civil War history, language, food, and conservatism, or resistance to change.

Ecotopia

Ecotopia, located along the northern part of the west coast of North America, is known for its natural beauty. People in Ecotopia are concerned about the environment in their region. Logging, mining, and other economic activities in Ecotopia rely on the region's natural resources.

The Empty Quarter

The Empty Quarter stretches over a large geographic area of the western United States and Canada where few people live, and large ranches dot the landscape. The United States government owns much of the land in the region. Some of the biggest corporations in the world are competing for the chance to develop it.

The Foundry

The Foundry includes the eastern Great Lakes area extending over to the Middle Atlantic region. It is an industrial region, with many of its cities, such as Detroit, in decline, following an era when they dominated world manufacturing. The Foundry is struggling to adjust to economic changes.

The Islands

The Islands is a a popular tourist destination, and includes the southern tip of Florida and the Caribbean Islands. Miami, Florida, has become the economic, cultural, and sometimes political center of the region due to its large Hispanic population.

MexAmerica

MexAmerica includes southern Texas, most of New Mexico and Arizona, and Southern California, along with Mexico. This is a region in which Hispanics vastly outnumber all other minorities. Radio, television, music, fashion, politics, food, language, and customs have been defined by the Spanish influence.

New England

New England, the part of the United States that English settlers first colonized, is the most easily defined of the "nine nations." This is the only one of Garreau's nations that acknowledges traditional political boundaries.

Quebec

Quebec is located in eastern Canada, and is the nation with a French-speaking population. Garreau states that this is the only independent nation in Canada.

Lesson 17
MAP SKILLS
Using Maps to Understand Different Ways of Defining Regions

The map below shows regions designated by their topography. If you compare this map with Garreau's "Nine Nations of North America" in the article, you will be able to compare two of the ways that North America can be regionalized.

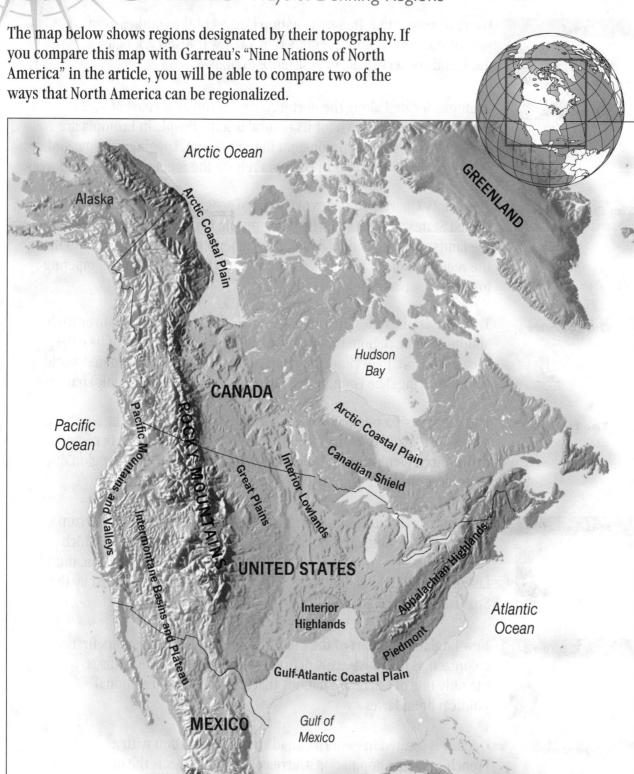

Arctic Ocean

Alaska

Arctic Coastal Plain

GREENLAND

Hudson Bay

CANADA

Arctic Coastal Plain

Canadian Shield

Pacific Ocean

Pacific Mountains and Valleys

ROCKY MOUNTAINS

Great Plains

Interior Lowlands

Appalachian Highlands

Atlantic Ocean

Intermontane Basins and Plateau

UNITED STATES

Interior Highlands

Piedmont

Gulf-Atlantic Coastal Plain

MEXICO

Gulf of Mexico

A. To help you make this comparison, complete the following.

1. Draw the regional boundaries shown on Garreau's map onto the topographical region map. Use a colored pen or pencil such as red or blue that will show up well.

2. Now write in the names Garreau gave to those regions. In this way, you will be **superimposing** maps, or putting one map on top of another.

B. Use the article map to help you locate and label the following cities on the superimposed map. Then answer the questions below.

1. Label the city of Winnipeg, Canada. To which regions can you say Winnipeg belongs?

   a. _____

   b. _____

2. Label the city of Toronto, Canada. To which regions can you say it belongs?

   a. _____

   b. _____

3. Label Miami, Florida. To which regions does it belong?

   a. _____

   b. _____

4. Label Los Angeles, California. To which regions does it belong?

   a. _____

   b. _____

C. Review your answers for section B above. Draw some conclusions about the conclusions about the regions identified.

1. The names of the regions originally shown on the superimposed map describe what?

   _____

2. The names of the regions from Joel Garreau's "nine nations" map describe what?

   _____

   _____

3. Now find a map in the Almanac showing different regions. What are the regions?

   _____

Lesson 17

ACTIVITY

Choose capitals for the
"Nine Nations of North America."

New Nations' Capitals

The **BIG**
Geographic Question

Which characteristics should be
considered in the selection of
capital cities for various regions?

From the article you learned about Joel Garreau's "nine nations" of North
America. The map skills lesson showed you an alternative way to regionalize
North America. Now select a capital city for each of Garreau's regions.

A. Collect information about the "nine nations" from the article. Copy the
below chart onto a sheet of paper and complete it. Use the Almanac to
help you identify cities of North America. One has been done for you.

"Nation"	Cities Included	Primary Language(s)	Physical Features	Major Economic Activities
Ecotopia	Vancouver, San Francisco	English	mountains, forests, coast	logging, mining, recreation
Empty Quarter				
Breadbasket				
Foundry				
MexAmerica				
Dixie				
Quebec				
New England				
Islands				

B. Imagine that each of these "nations" needs to choose a capital city. Use what you have learned about them to make an appropriate selection. Write your choices on the lines below, and support each choice with reasons why you selected it. For instance, you might choose Cleveland for the capital of The Foundry because it is centrally located and is an important manufacturing and shipping center on Lake Erie with access to the St. Lawrence Seaway and the Atlantic Ocean.

Ecotopia _____

_____

The Empty Quarter _____

_____

The Breadbasket _____

_____

The Foundry _____

_____

MexAmerica _____

_____

Dixie _____

_____

Quebec _____

_____

New England _____

_____

The Islands _____

_____

Connecting Oceans

As you read about the Panama Canal, think about how people overcame natural obstacles in order to make travel faster and easier.

Geography plays a major role in how goods are transported from one area to another. Physical features such as mountains can make transportation difficult. To make transport by water easier, people have built **canals.** Canals are constructed waterways through which ships can travel. They are typically built across an **isthmus,** a narrow strip of land that joins two larger bodies of land and that has water on both sides. The Panama Canal is one example.

When looking at a map that shows North America and South America, it is easy to see why the Panama Canal was necessary. Prior to 1914, when the canal was built, in order for a ship to travel from New York City to San Francisco, it had to travel around South America. The trip took about four months. South America is 4,700 miles long from the Caribbean Sea at its northern tip to its southern tip near Antarctica.

Travelers could not easily stop to rest in South America, especially along its south and west coasts. To the south, the waters of the Pacific Ocean were very dangerous. Storms caught many sailors by surprise, tossing and often sinking their ships. To the west, the Andes Mountains made it difficult to settle inland. Cities where sailors could easily dock for rest and supplies were hard to find. A faster, more efficient route was built in Panama.

There were problems that had to be solved before the canal could be completed. The water in the Atlantic and Pacific oceans is at different levels. Workers built three sets of chambers in the canal called *locks* where water levels could be raised or lowered. The locks allowed the ships to move more easily across the isthmus from one ocean to the other.

Here's how a lock works. Once a ship enters a lock, huge steel gates close behind it. Water is let into or out of the chamber from below until the water in the chamber reaches the level of the next body of water the ship must enter. Then the gates in front of the ship open, and it moves on. It takes a ship about eight hours to pass through the Panama Canal and move from one ocean to the other. Once the canal was complete, that same trip from New York to San Francisco took about 47 days. To make travel even more efficient, the locks were built in pairs so that ships traveling in opposite directions can use the canal at the same time.

Aerial view of Panama Canal

It took more than 30 years to accomplish the enormous engineering feat that is the Panama Canal. Tens of thousands of people worked on it. Today about 38 vessels pass through the canal every day, most of them carrying thousands of tons of cargo. As one can see, geography was an important factor in the development of the Panama Canal. A physical feature, an isthmus, helped to solve the problem of transporting goods from one ocean to another. The Panama Canal is an excellent example of people using the geography of a region to improve their lives.

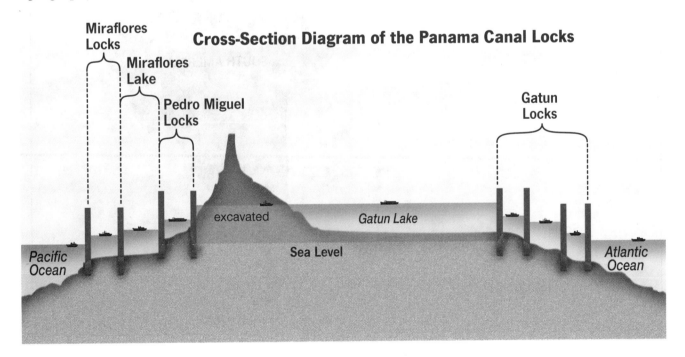

Cross-Section Diagram of the Panama Canal Locks

Miraflores Locks

Miraflores Lake

Pedro Miguel Locks

Gatun Locks

excavated

Gatun Lake

Pacific Ocean

Sea Level

Atlantic Ocean

Lesson 18

MAP SKILLS
Using a Map to See How Technological Advances Affect Travel Time

In the mid-1800s many ships traveled between New York City and San Francisco by sailing around Cape Horn at the southern tip of South America. A one-way journey took at least four months. With the development of steam power, ships could travel much faster. Technology seemed to shrink the distance. In 1869 the United States' transcontinental railroad was completed. People and materials could travel from coast to coast by train. But large, heavy cargoes still had to travel by ship.

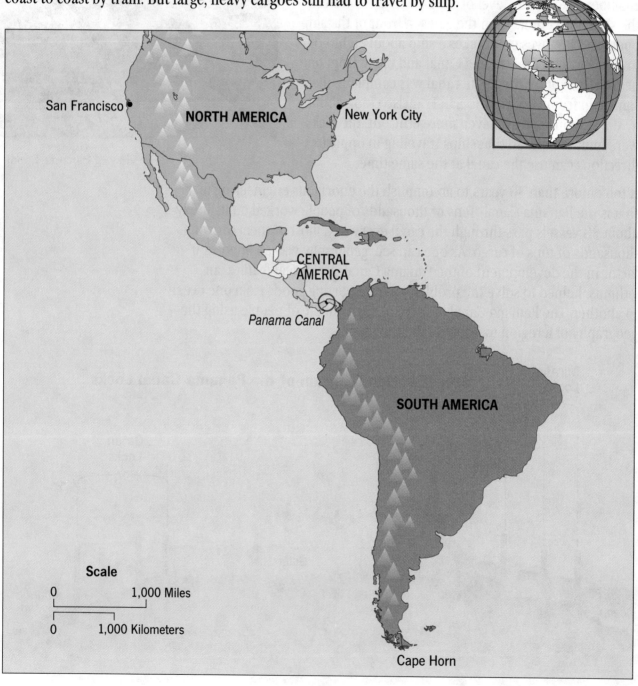

San Francisco

NORTH AMERICA

New York City

CENTRAL AMERICA

Panama Canal

SOUTH AMERICA

Scale

0   1,000 Miles

0   1,000 Kilometers

Cape Horn

A. Draw the following ocean routes between New York City and San Francisco on the map.

   1. The shortest route going around Cape Horn at the southern tip of South America

   2. The shortest route via the Panama Canal

B. Figure out the approximate distance and travel time for the routes you drew on the map.

   1. Use a piece or string and a ruler to measure the length of each route in inches.

   a. The route via Cape Horn is about _____ inches long.

   b. The route via the Panama Canal is about _____ inches long.

   2. Now use the map scale to calculate the approximate number of miles in each route.

   (Hint: Multiply the number of inches in the route times the number of miles per inch to get the total length of the route.)

   a. The route via Cape Horn is about _____ miles long.

   b. The route via the Panama Canal is about _____ miles long.

   3. The *Mauretania,* one of the fastest ocean-going steamships in the early 1900s, could travel about 745 miles per day. About how many days would it take the *Mauretania* to travel each route from New York City to San Francisco?

   (Hint: Divide the length of route in miles by the number of miles per day to get the number of days needed to travel the route.)

   a. The voyage via Cape Horn would take about _____ days.

   b. The voyage via the Panama Canal would take about _____ days.

   4. How many days would the ship save by using the Panama Canal?

   _____

   5. Based on what you have learned from the above, write a couple of sentences explaining how the trip from New York City to San Francisco was affected by technological advances. Be sure to describe what those technological advances were.

   _____

   _____

   _____

   _____

Lesson 18

 Choose a good location for a new Central American canal.

Planning a New Canal

The **BIG** Geographic Question

What would be a good site for a new canal joining the Atlantic and Pacific oceans?

From the article you learned about both the advantages and disadvantages of the site for the Panama Canal. In the map skills lesson you saw how the Panama Canal changed the travel distance and time of trips by ship from the east to west coasts of North America. Now plan the site of a new canal in another location on the Central America isthmus.

A. Study the Central American map on page 116 in the Almanac. Think about the features of each country in Central America that might make it a good place or a poor place for a new canal to meet the world's shipping needs in the twenty-first century.

1. Which countries have the narrowest widths?

_____

_____

2. Which countries have the narrowest mountain ranges?

_____

_____

3. Why do you think Belize and El Salvador are not possibilities for building a new canal in Central America?

_____

_____

_____

_____

_____

108

B. Use the Almanac and other research materials to find out more about the five countries that have coasts on both oceans. In the chart below list what you think would be the advantages and disadvantages of each country as a site for a new canal. What physical features do you see on the map that would help a country or put it at a disadvantage in the building of a canal?

Country	Advantages	Disadvantages
Guatemala		
Honduras		
Nicaragua		
Costa Rica		
Panama		

C. Which country would you choose as the best site for a new canal? State your preference and support it with reasons.

_____

_____

_____

_____

_____

_____

ALMANAC

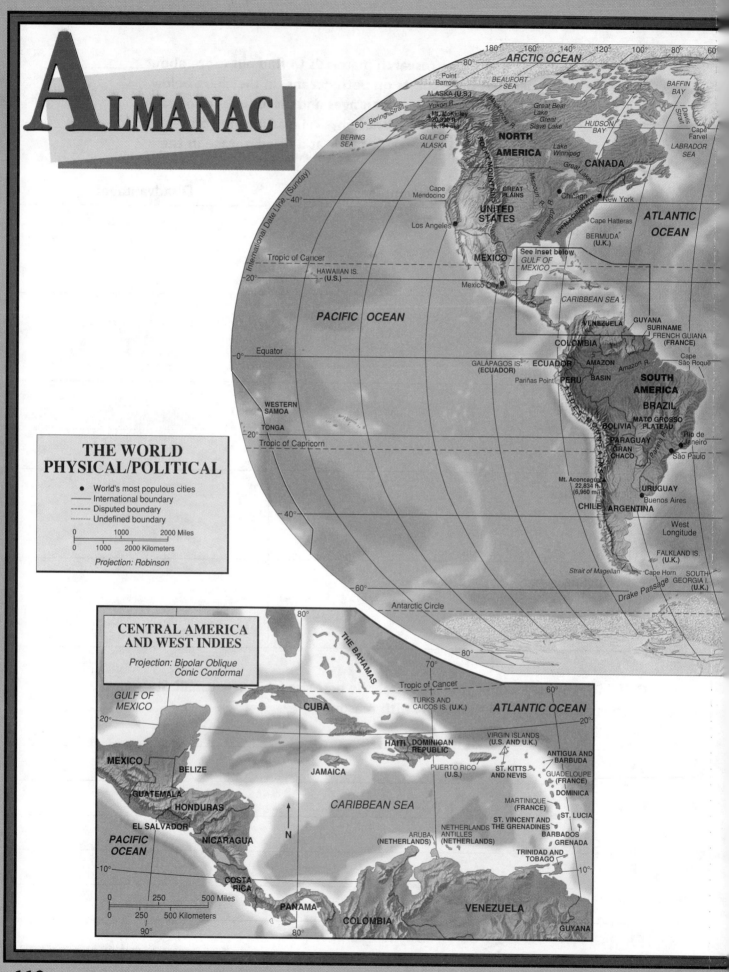

THE WORLD PHYSICAL/POLITICAL

- • World's most populous cities
- — International boundary
- --- Disputed boundary
- ······ Undefined boundary

| 0 | 1000 | 2000 Miles |
| 0 | 1000 | 2000 Kilometers |

Projection: Robinson

CENTRAL AMERICA AND WEST INDIES

Projection: Bipolar Oblique Conic Conformal

| 0 | 250 | 500 Miles |
| 0 | 250 | 500 Kilometers |

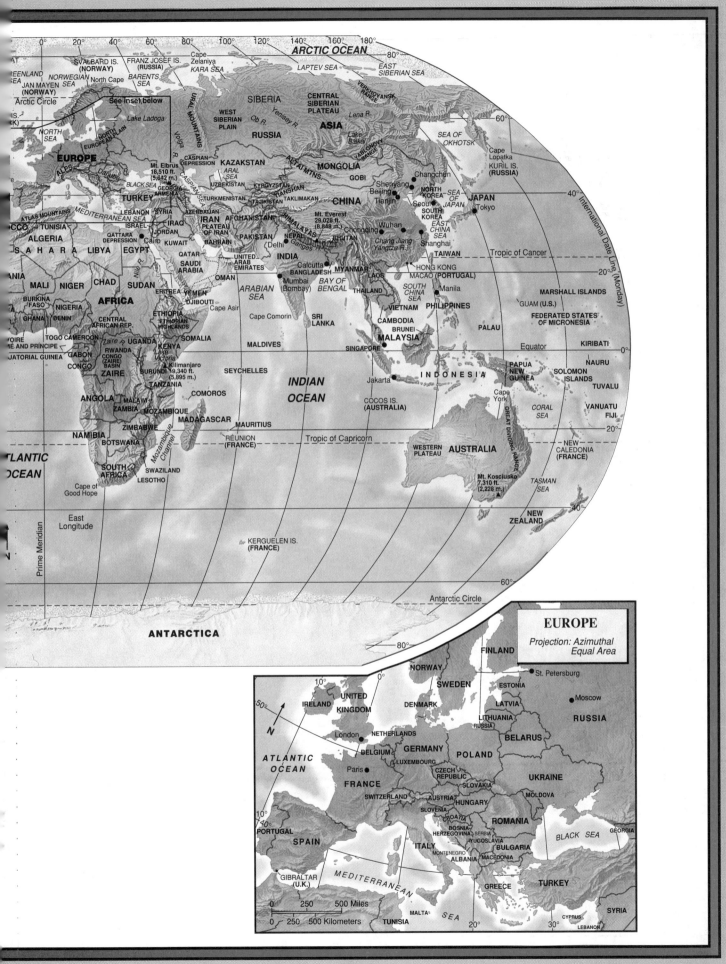

ARCTIC OCEAN

Cape Zelaniya
SVALBARD IS. (NORWAY)
FRANZ JOSEF IS. (RUSSIA)
KARA SEA
LAPTEV SEA
EAST SIBERIAN SEA
80°
GREENLAND
JAN MAYEN (NORWAY)
NORWEGIAN SEA
North Cape
BARENTS SEA
VERKHOYANSK RANGE
Arctic Circle
NORTH SEA
See Inset below
URAL MOUNTAINS
SIBERIA
CENTRAL SIBERIAN PLATEAU
60°
EUROPE
Lake Ladoga
WEST SIBERIAN PLAIN
Yenisey R.
Lena R.
Lake Baikal
SEA OF OKHOTSK
ALPS
EUROPEAN PLAIN
Volga R.
Ob R.
ASIA
RUSSIA
Cape Lopatka
KURIL IS. (RUSSIA)
Danube
Mt. Elbrus 18,510 ft. (5,642 m.)
CASPIAN DEPRESSION
KAZAKSTAN
ARAL SEA
MONGOLIA
ALTAI MTNS.
GOBI
Changchen
YABLONOY RANGE
40°
TURKEY
BLACK SEA
GEORGIA
ARMENIA
CASPIAN SEA
UZBEKISTAN
TURKMENISTAN
KYRGYZSTAN
TIANSHAN
TAKLIMAKAN
Shenyang
Beijing
NORTH KOREA
Seoul
SOUTH KOREA
SEA OF JAPAN
JAPAN
Tokyo
ATLAS MOUNTAINS
MEDITERRANEAN SEA
LEBANON
SYRIA
ISRAEL
IRAQ
JORDAN
AZERBAIJAN
TAJIKISTAN
AFGHANISTAN
IRAN
PLATEAU OF IRAN
PAKISTAN
HIMALAYAS
Mt. Everest 29,028 ft. (8,848 m.)
NEPAL
BHUTAN
Tianjin
Wuhan
Chongqing
Chang Jiang Yangtze R.
Shanghai
EAST CHINA SEA
MOROCCO
TUNISIA
ALGERIA
LIBYA
EGYPT
KUWAIT
BAHRAIN
QATAR
SAUDI ARABIA
UNITED ARAB EMIRATES
OMAN
Delhi
Ganges R.
INDIA
Calcutta
BANGLADESH
MYANMAR
LAOS
CHINA
TAIWAN
HONG KONG
MACAO (PORTUGAL)
Tropic of Cancer
20°
SAHARA
Nile R.
Cairo
QATTARA DEPRESSION
SUDAN
ERITREA
YEMEN
DJIBOUTI
Cape Asir
ARABIAN SEA
Mumbai (Bombay)
BAY OF BENGAL
THAILAND
VIETNAM
CAMBODIA
Manila
SOUTH CHINA SEA
MARSHALL ISLANDS
GUAM (U.S.)
MAURITANIA
MALI
NIGER
CHAD
AFRICA
ETHIOPIA
ETHIOPIAN HIGHLANDS
SOMALIA
Cape Comorin
SRI LANKA
MALDIVES
BRUNEI
MALAYSIA
PHILIPPINES
PALAU
FEDERATED STATES OF MICRONESIA
BURKINA FASO
NIGERIA
BENIN
CENTRAL AFRICAN REP.
UGANDA
KENYA
Lake Victoria
SINGAPORE
Equator
KIRIBATI
0°
GHANA
TOGO
CAMEROON
GABON
CONGO
ZAIRE
CONGO (ZAIRE) BASIN
RWANDA
BURUNDI
Kilimanjaro 19,340 ft. (5,895 m.)
TANZANIA
SEYCHELLES
INDIAN OCEAN
Jakarta
INDONESIA
PAPUA NEW GUINEA
SOLOMON ISLANDS
NAURU
TUVALU
IVORY COAST
SÃO TOMÉ AND PRÍNCIPE
EQUATORIAL GUINEA
ANGOLA
MALAWI
ZAMBIA
MOZAMBIQUE
MADAGASCAR
COMOROS
MAURITIUS
COCOS IS. (AUSTRALIA)
Cape York
CORAL SEA
VANUATU
FIJI
NAMIBIA
BOTSWANA
ZIMBABWE
SWAZILAND
LESOTHO
Mozambique Channel
RÉUNION (FRANCE)
Tropic of Capricorn
GREAT DIVIDING RANGE
NEW CALEDONIA (FRANCE)
20°
ATLANTIC OCEAN
SOUTH AFRICA
Cape of Good Hope
WESTERN PLATEAU
AUSTRALIA
TASMAN SEA
Prime Meridian
East Longitude
Mt. Kosciusko 7,310 ft. (2,228 m.)
NEW ZEALAND
40°
KERGUELEN IS. (FRANCE)
International Date Line (Monday)
Antarctic Circle
60°
ANTARCTICA
80°

EUROPE
Projection: Azimuthal Equal Area

FINLAND
NORWAY
SWEDEN
ESTONIA
St. Petersburg
LATVIA
Moscow
RUSSIA
IRELAND
UNITED KINGDOM
DENMARK
LITHUANIA
RUSSIA
BELARUS
London
NETHERLANDS
GERMANY
POLAND
ATLANTIC OCEAN
BELGIUM
LUXEMBOURG
CZECH REPUBLIC
UKRAINE
Paris
FRANCE
SWITZERLAND
SLOVAKIA
AUSTRIA
HUNGARY
MOLDOVA
SLOVENIA
CROATIA
ROMANIA
PORTUGAL
SPAIN
ITALY
BOSNIA HERZEGOVINA
SERBIA
YUGOSLAVIA
MONTENEGRO
ALBANIA
MACEDONIA
BULGARIA
BLACK SEA
GEORGIA
GIBRALTAR (U.K.)
MEDITERRANEAN SEA
GREECE
TURKEY
MALTA
TUNISIA
CYPRUS
SYRIA
LEBANON
N
10°
0°
10°
20°
30°
40°
50°

0   250   500 Miles
0   250   500 Kilometers

111

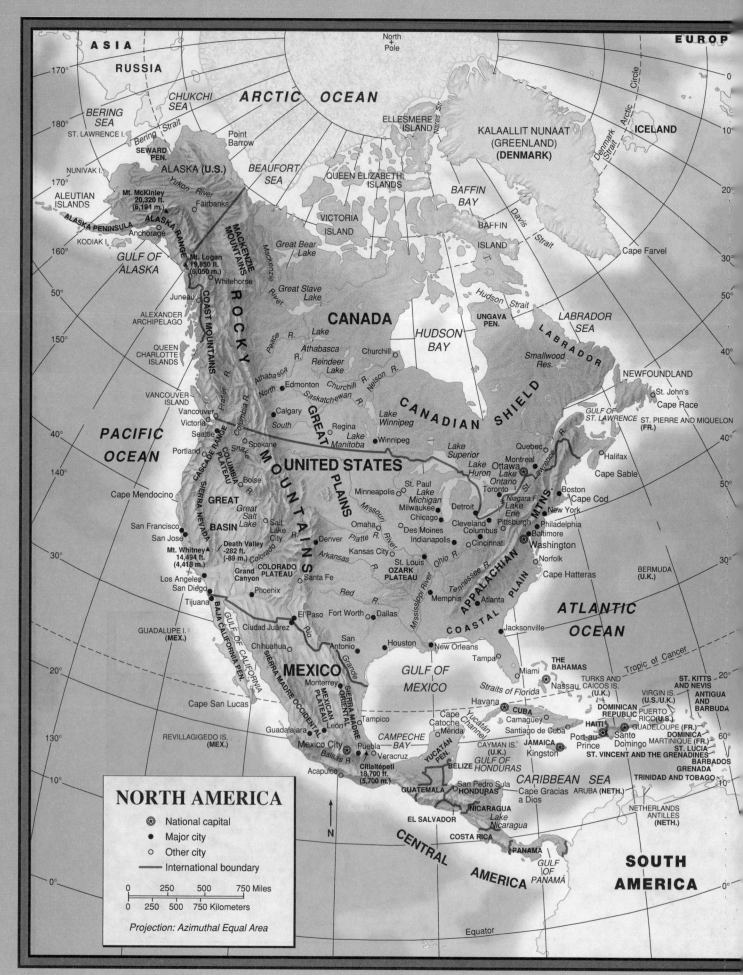

ASIA
RUSSIA
North Pole
EUROP
ICELAND
ARCTIC OCEAN
170°
CHUKCHI SEA
BERING SEA
ST. LAWRENCE I.
Bering Strait
Point Barrow
ELLESMERE ISLAND
KALAALLIT NUNAAT (GREENLAND) (DENMARK)
Arctic Circle
180°
0°
10°
20°
SEWARD PEN.
NUNIVAK I.
ALASKA (U.S.)
BEAUFORT SEA
QUEEN ELIZABETH ISLANDS
BAFFIN BAY
Denmark Strait
ALEUTIAN ISLANDS
Mt. McKinley 20,320 ft. (6,191 m.)
Fairbanks
Yukon River
VICTORIA ISLAND
BAFFIN ISLAND
Davis Strait
170°
ALASKA PENINSULA
ALASKA RANGE
Anchorage
MACKENZIE MOUNTAINS
Cape Farvel
30°
KODIAK I.
Mt. Logan 19,850 ft. (6,050 m.)
Whitehorse
Mackenzie River
GULF OF ALASKA
Juneau
Great Bear Lake
Hudson Strait
160°
Great Slave Lake
UNGAVA PEN.
LABRADOR SEA
40°
ALEXANDER ARCHIPELAGO
COAST MOUNTAINS
CANADA
Peace R.
Lake Athabasca
Reindeer Lake
Churchill
HUDSON BAY
LABRADOR
Smallwood Res.
NEWFOUNDLAND
150°
QUEEN CHARLOTTE ISLANDS
Athabasca R.
North
Edmonton
Churchill R.
Nelson R.
CANADIAN SHIELD
GULF OF ST. LAWRENCE
St. John's
Cape Race
VANCOUVER ISLAND
Fraser R.
ROCKY
Calgary
Saskatchewan R.
South
Regina
Lake Winnipeg
Lake Manitoba
St. Lawrence R.
ST. PIERRE AND MIQUELON (FR.)
140°
Vancouver
Victoria
Seattle
Columbia R.
GREAT
Spokane
Winnipeg
Lake Superior
Quebec
Montreal
Halifax
Cape Sable
40°
PACIFIC OCEAN
Portland
Snake R.
Boise
MOUNTAINS
UNITED STATES
PLAINS
Minneapolis
St. Paul
Milwaukee
Lake Michigan
Lake Huron
Ottawa
Lake Ontario
Toronto
Lake Erie
Niagara Falls
APPALACHIAN MTNS.
Boston
Cape Cod
New York
50°
Cape Mendocino
CASCADE RANGE
COLUMBIA PLATEAU
GREAT BASIN
Great Salt Lake
Salt Lake City
Missouri R.
Omaha
Chicago
Des Moines
Detroit
Cleveland
Columbus
Cincinnati
Pittsburgh
Philadelphia
Baltimore
Washington
San Francisco
San Jose
SIERRA NEVADA
Mt. Whitney 14,494 ft. (4,418 m.)
Death Valley -282 ft. (-89 m.)
Denver
Platte R.
Colorado R.
Kansas City
St. Louis
Ohio R.
Norfolk
Cape Hatteras
BERMUDA (U.K.)
30°
Los Angeles
San Diego
Tijuana
COLORADO PLATEAU
Grand Canyon
Santa Fe
Arkansas R.
Red R.
OZARK PLATEAU
Tennessee R.
Mississippi River
Memphis
Atlanta
COASTAL PLAIN
ATLANTIC OCEAN
30°
GUADALUPE I. (MEX.)
GULF OF CALIFORNIA
BAJA CALIFORNIA PEN.
Phoenix
El Paso
Ciudad Juárez
Fort Worth
Dallas
Houston
New Orleans
Jacksonville
Tampa
San Antonio
Rio Grande
MEXICO
Chihuahua
SIERRA MADRE OCCIDENTAL
MEXICAN PLATEAU
SIERRA MADRE ORIENTAL
GULF OF MEXICO
Miami
THE BAHAMAS
Nassau
TURKS AND CAICOS IS. (U.K.)
Tropic of Cancer
ST. KITTS AND NEVIS
20°
Cape San Lucas
Monterrey
León
Tampico
Straits of Florida
Havana
CUBA
Camaguey
Santiago de Cuba
VIRGIN IS. (U.S./U.K.)
PUERTO RICO (U.S.)
DOMINICAN REPUBLIC
ANTIGUA AND BARBUDA
GUADELOUPE (FR.)
130°
REVILLAGIGEDO IS. (MEX.)
Guadalajara
Yucatán Channel
Cape Catoche
Mérida
CAYMAN IS. (U.K.)
JAMAICA
Kingston
HAITI
Port-au-Prince
Santo Domingo
DOMINICA
MARTINIQUE (FR.)
ST. LUCIA
60°
Mexico City
Puebla
Veracruz
Balsas R.
YUCATÁN PEN.
BELIZE
GULF OF HONDURAS
CARIBBEAN SEA
ST. VINCENT AND THE GRENADINES
BARBADOS
GRENADA
TRINIDAD AND TOBAGO
Acapulco
Citlaltépeti 18,700 ft. (5,700 m.)
CAMPECHE BAY
San Pedro Sula
GUATEMALA
HONDURAS
Cape Gracias a Dios
ARUBA (NETH.)
10°
EL SALVADOR
NICARAGUA
Lake Nicaragua
NETHERLANDS ANTILLES (NETH.)
COSTA RICA
SOUTH AMERICA
CENTRAL AMERICA
PANAMA
GULF OF PANAMÁ

NORTH AMERICA

⊛ National capital
● Major city
○ Other city
— International boundary

0   250   500   750 Miles
0   250 500 750 Kilometers

Projection: Azimuthal Equal Area

N

Equator
0°

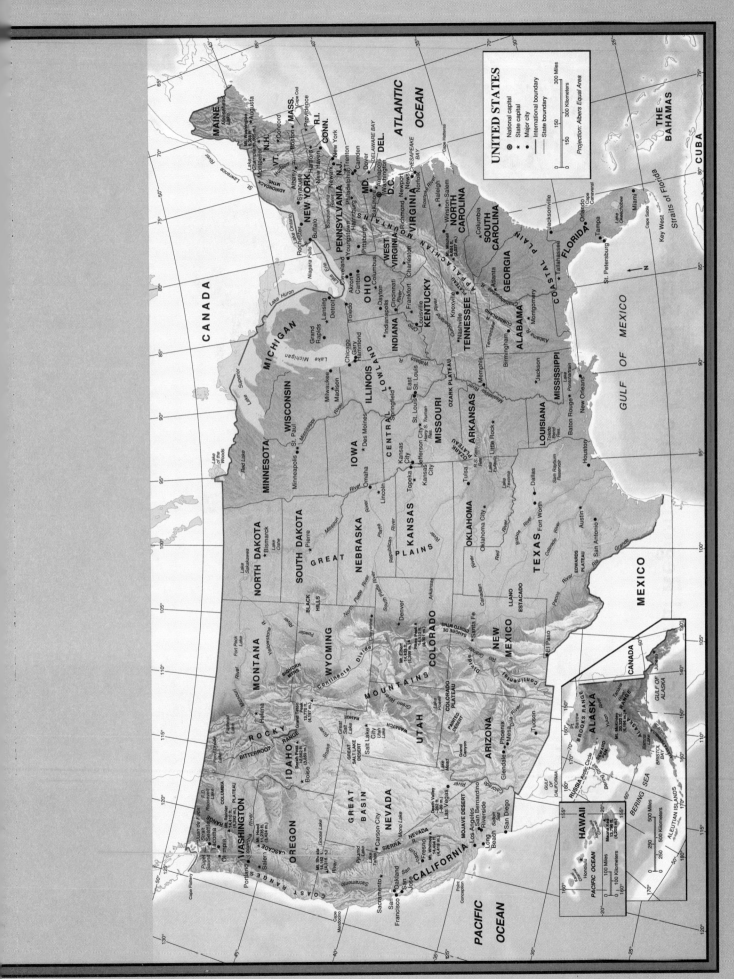

UNITED STATES

◉ National capital
★ State capital
● Major city
— International boundary
— State boundary

Projection: Albers Equal Area

300 Miles
300 Kilometers
150
0

Agricultural Regions of the United States

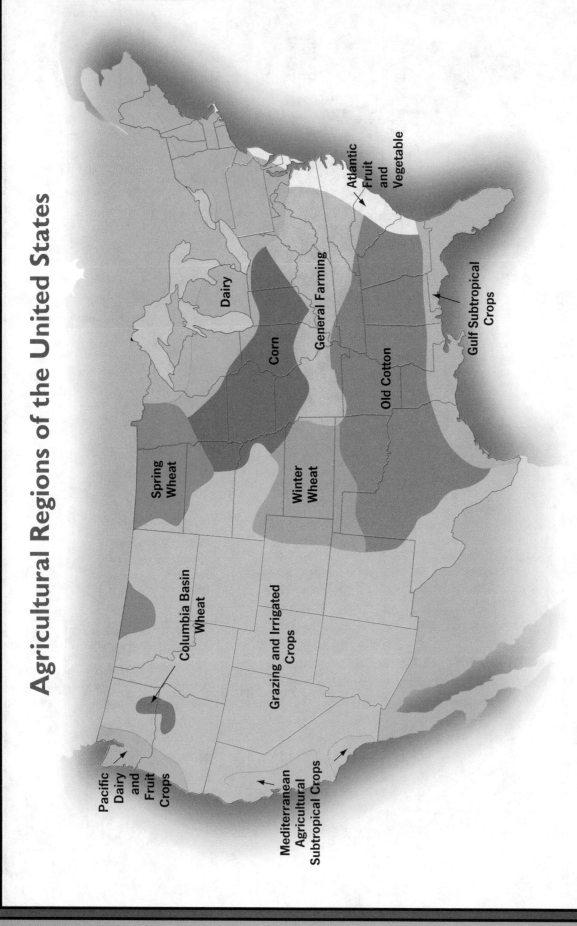

- Atlantic Fruit and Vegetable
- Dairy
- General Farming
- Gulf Subtropical Crops
- Corn
- Old Cotton
- Spring Wheat
- Winter Wheat
- Columbia Basin Wheat
- Grazing and Irrigated Crops
- Pacific Dairy and Fruit Crops
- Mediterranean Agricultural Subtropical Crops

Hurricanes, Typhoons, and Cyclones

A hurricane is a storm with violent winds of more than 75 miles per hour circulating around a calm center. Hurricanes are usually accompanied by heavy rains, high tides, and floods, in regions along the coasts.

This type of storm is called a hurricane when it occurs in the eastern Pacific and Atlantic Oceans, a typhoon when it occurs in the China Sea and the western Pacific Ocean, and a tropical cyclone when it occurs in the Indian Ocean. Winds circulate in a counterclockwise direction in the northern hemisphere, and clockwise in the southern hemisphere.

Hurricanes, typhoons, and cyclones form over warm tropical and subtropical waters during the time of year when water temperatures are the warmest, and humidity is the highest. These storms form when an easterly wave, an area of low pressure, deepens and intensifies. The warm water and air over these low pressure areas create instability. Winds begin circulating and the storm moves westward, growing stronger and larger. Some hurricanes measure 200–300 miles across. Winds swirl around an eye, a 20-mile diameter calm spot in the center of the storm. Wall clouds are clouds surrounding the eye. They contain the strongest winds and heaviest rain.

Hurricane winds can reach speeds of 130 to 150 miles per hour. High speed winds, combined with the force of the ocean produce huge waves called storm surges, which cause flooding. Heavy rainfall caused by the storm also causes flooding. Tornadoes are

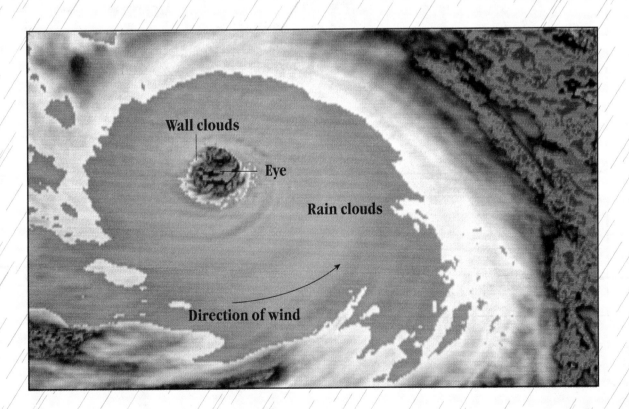

Wall clouds

Eye

Rain clouds

Direction of wind

Source: *World Book Encyclopedia*, 1991;
Modern Physical Geography, Alan H. and Arthur N. Strahler, 1992

Central America

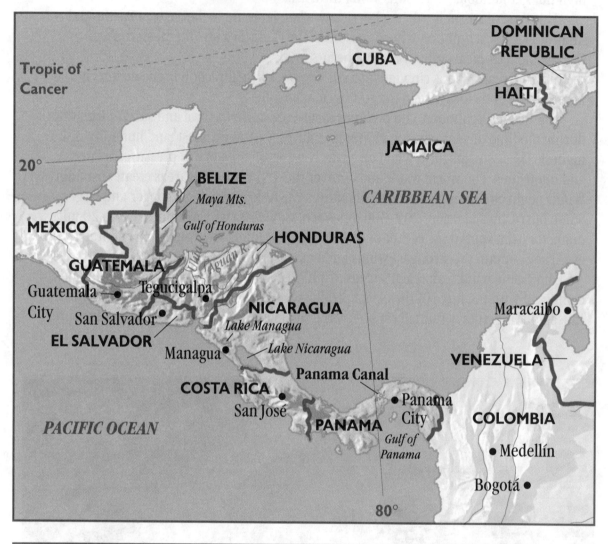

Tropic of Cancer

20°

CUBA

DOMINICAN REPUBLIC

HAITI

JAMAICA

CARIBBEAN SEA

BELIZE

Maya Mts.

Gulf of Honduras

MEXICO

HONDURAS

GUATEMALA

Ulúa R.

Aguán R.

Guatemala City

Tegucigalpa

NICARAGUA

San Salvador

Lake Managua

EL SALVADOR

Managua

Lake Nicaragua

Maracaibo

VENEZUELA

Panama Canal

COSTA RICA

Panama City

COLOMBIA

San José

PANAMA

Gulf of Panama

Medellín

PACIFIC OCEAN

Bogotá

80°

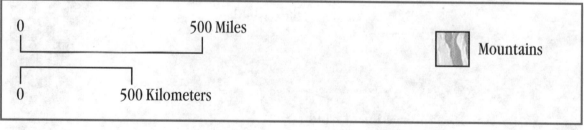

0　　　　　　　500 Miles

0　　　500 Kilometers

Mountains

Title: _____

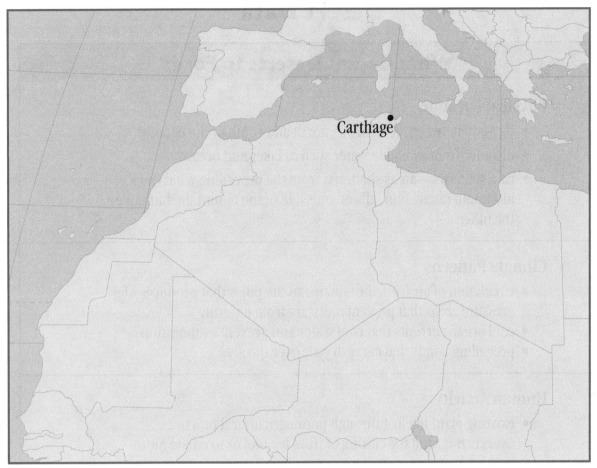

Carthage

Key

Resources of the Carthaginian Empire

Carthage was located on a penisula in North Africa near the present-day city of Tunis. It had a strong base in trade across the Sahara Desert and over water through its excellent harbors.

Carthaginians eagerly explored the region for resources. They found and exploited silver mines in the Sahara Desert of North Africa and in southern Spain. They also obtained tin, silver, gold, and iron from ethnic groups in the region, in exchange for consumer goods. The forested mountains provided a supply of wood and wild game.

Agricultural products were also important resources for the Carthaginian Empire. The foothills of the Atlas Mountains provided adequate soil for extensive vineyards. Grains such as wheat were grown on the coastal plains.

Desert Data

What Causes Deserts to Form

Physical Location

- location between 15° and 30° north and south of the equator
- distance from available water such as lakes and oceans
- rain shadows — areas sheltered from the prevailing winds by a mountain range; rain falls on one side of the mountain, but not on the other

Climate Patterns

- circulation of air from the equator to the poles that produces high pressure areas that prevent moisture from entering
- cold ocean currents that cool water and prevent evaporation
- prevailing winds that carry dry air over the area

Human Activity

- existing plant life lost through poor agricultural practices, overgrazing, and the cutting of trees for fuel or to create more cropland
- mining that strips the land and contaminates the water
- inefficient water management
- deep wells that deplete fossil water which cannot be replaced

Desert Data

The Kalahari Desert

The Kalahari Desert is not considered a true desert by some scientists who claim that a true desert receives less than ten inches of rain per year. Many scientists argue that the Kalahari is a semidesert since it receives as much as 15 inches of rainfall per year in some places. Scientists say that it is not the lack of rainfall that makes this area a desert. Instead, they explain that the sandy soils and lack of surface water, such as rivers and lakes, are the reasons for the lack of lush vegetation.

Month	Average Monthly Rainfall (inches)
January	4.2
February	3.1
March	2.8
April	0.7
May	0.2
June	0.1
July	0
August	0
September	0
October	0.9
November	2.2
December	3.4

The Sahara Desert

The Sahara Desert stretches across northern Africa and covers parts of ten countries, making it the world's largest desert. It is also the world's driest desert. Almost all parts of the Sahara receive less than ten inches of rainfall each year, making it barren.

Month	Average Monthly Rainfall (inches)
January	0.1
February	0.1
March	0
April	0
May	0
June	0
July	0.1
August	0
September	0
October	0
November	0.2
December	0.1

Sources: *World Weather Guide*, E.A. Pearce and Gordon Smith, 1990; *Grolier Electronic Encyclopedia*, 1995

Example of a Food Web

A food web shows all of the feeding relationships within an ecosystem.

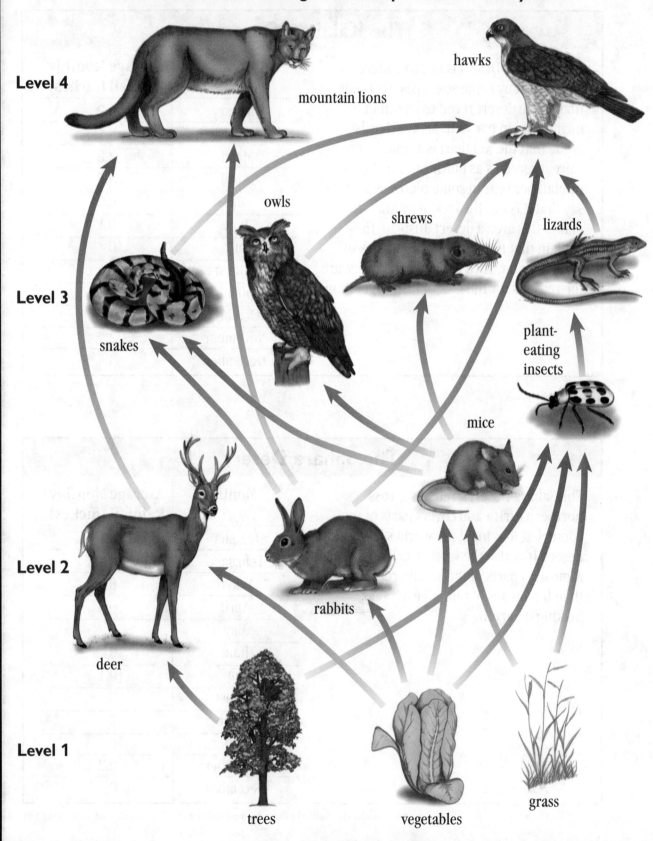

Level 4

mountain lions

hawks

Level 3

owls

shrews

lizards

snakes

plant-eating insects

mice

Level 2

rabbits

deer

Level 1

trees

vegetables

grass

Animals Found in the Ecosystem Around Lake Victoria

• crocodiles	• microscopic algae
• humans	• brine shrimp
• cichlids	• insect larvae
• Nile perch	

The Effect of Human Action on the Ecology of Lake Victoria

The decrease in the numbers of cichlids in Lake Victoria was only one effect of human activities upon the environment. Here are some others:

- Early in the twentieth century Europeans cut down forests in the Lake Victoria watershed to plant coffee and tea plantations.

- Kisumu, Kenya, is surrounded by tropical grassland called a **savanna**. Only a few trees grow in this grassland. Many of these trees have been cut down and burned because the easiest way to preserve Nile perch is drying them over a wood fire.

- Sewage from livestock operations and sugar refining plants runs into the lake.

- Fertilizer from surrounding farmland has washed into the lake, causing rapid algae growth. Decaying algae uses the oxygen in the lake, leaving little for fish to breathe. Some species of cichlids that ate algae have disappeared. Since then, there has been nothing to stop the growth of algae.

Major Straits of the World

Strait	Size	Control	History
Dardanelles (Sea of Marmara to the Aegean Sea)	5 miles wide; 1 mile wide at its narrowest point; 38 miles long; 200 feet deep	Turkey	This strait controls access to the cities northeast of Turkey. It has been of great strategic and economic importance throughout history. During World War I Allied fleets attempted to take control of the strait. The Turks held the Allies off, securing control of the area.
Bosporus (Black Sea to the Sea of Marmara)	2.3 miles wide; 800 yards wide at its narrowest point; 19 miles long	Turkey	Pressure from Russia and Europe resulted in several treaties, leading to the Montreux Convention of 1936. Turkey was granted full control of the strait.
Hormuz (Persian Gulf to the Gulf of Oman)	40–60 miles wide	No country has been given control of the strait.	This strait is an important petroleum shipping route from the oil-producing countries of southwest Asia.
Gibraltar (Atlantic Ocean to the Mediterranean Sea)	8–23 miles wide; 32 miles long; 1,050 feet deep at its shallowest point	No country has been given control of the strait.	Control of this strategic strait has long been valued because it permits entry into the Mediterranean, and therefore, access to Europe, Asia, and Africa. The strait is also important because as surface water moves east from the Atlantic, the deeper water moves west toward the Atlantic. This movement helps circulate the Mediterranean Sea.
Suez Canal* (Mediterranean Sea to the Red Sea) * Functions like a strait; is really an isthmus	741 feet wide at surface and 302 wide at the bottom; 118 miles long; 64 feet deep	Egypt	In 1858 France and Egypt agreed to build a maritime canal, creating the Suez Canal Company which was to operate the canal for 99 years. In 1875 Britain bought out Egypt's share of the canal. In 1956 Egypt took control of the company, and seized all of its Egyptian assets.

Source: *World Book Encyclopedia*, 1991; *National Geographic Exploring Your World*, 1993; *National Geographic Picture Atlas of Our World*, 1993

Europe's Rivers

River	Length	Drainage Area	Location
Volga	2,193 mi.	533,000 sq. mi.	Russia; northwest of Moscow, flows to the landlocked Caspian Sea
Danube	1,770 mi.	315,000 sq. mi.	Germany; Black Forest Mountains; flows to the Black Sea
Dnieper	1,367 mi.	195,000 sq. mi.	Ukraine; through Kiev; flows to the Black Sea
Rhine	820 mi.	86,700 sq. mi.	begins in Swiss Alps; flows through Germany and Netherlands to Rotterdam (the Netherlands' leading seaport) to the North Sea
Elbe	724 mi.	55,620 sq. mi.	begins in Krkonose Mountains in the Czech Republic; flows through Germany to the North Sea
Rhône	505 mi.	37,750 sq. mi.	begins in Swiss Alps; flows through Switzerland and France to the Mediterranean Sea

More About Four Rivers of Europe

The Volga and Dnieper rivers are located in eastern Europe. The Volga River is Europe's longest river. It begins in the hills northwest of Moscow and flows through many cities on its way to the Caspian Sea. It receives water from some 200 tributaries, which makes it Russia's most important internal waterway. The Dnieper River is the third longest river in Europe and is a vital trade route. It flows through regions of heavy deposits of coal, iron, and natural gas in the Ukraine. The presence of these natural resources has resulted in the growth of industry along the Dnieper River.

Located in central Europe, the Danube River is the second longest river in Europe. It begins in the Black Forest region of Germany and flows into the Black Sea. Its course passes through nine countries and includes almost 300 tributaries. The Danube flows through less-developed, mostly agricultural areas of Europe. It has long been noted for its beauty, and its banks serve as the site of some of Europe's greatest cities, including Vienna, Budapest, and Belgrade.

The Elbe River begins in the mountains of the Czech Republic and flows through Germany to the North Sea. Two great cities along its banks are Hamburg and Dresden.

Sources: *Geography: Realms, Regions, and Concepts*, H.J. De Blij and P. Muller, 1994; *Essentials of World Regional Geography*, J.H. Wheeler, Jr. and J.T. Kostbade, 1993

Population Density of Europe

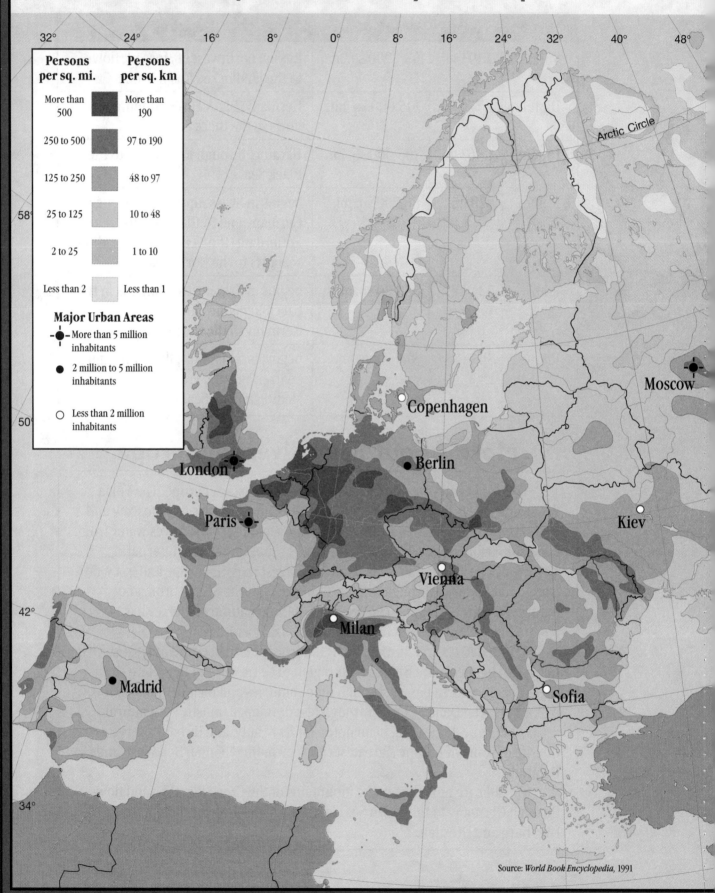

Persons per sq. mi. | Persons per sq. km
More than 500 | More than 190
250 to 500 | 97 to 190
125 to 250 | 48 to 97
25 to 125 | 10 to 48
2 to 25 | 1 to 10
Less than 2 | Less than 1

Major Urban Areas
More than 5 million inhabitants
2 million to 5 million inhabitants
Less than 2 million inhabitants

Arctic Circle

Moscow
Copenhagen
Berlin
London
Kiev
Paris
Vienna
Milan
Madrid
Sofia

Source: *World Book Encyclopedia*, 1991

Asia's Four "Little Tigers"

Indicators of Economic Growth

	Hourly Wage	Growth Rate	Unemployment Rate
Singapore	$5.25	9.9%	1.9%
Hong Kong	$4.29	5.2%	2.3%
South Korea	$5.53	8.4%	2.4%
Taiwan	$5.22	6.4%	1.6%

Trade	Singapore	Hong Kong	South Korea	Taiwan
Trading Partners	U.S., European Union, Hong Kong, Japan	U.S., United Kingdom, Japan, Germany, China, Taiwan	U.S., Japan	U.S., Hong Kong, Japan, Germany
Major Resources	poultry, vegetables, fruit, orchids	rice, vegetables, livestock	grains, vegetables, fruit	sugarcane, rice, vegetables, fruit
Imported Goods and Their Value	aircraft, petroleum, chemicals, food; total value = $66.4 billion	raw materials, transport equipment, food; total value = $149.6 billion	chemicals, textiles, steel, oil, grains, chemicals, machinery, electronics; total value = $79.1 billion	machinery, basic metals, crude oil, chemicals; total value = $77.1 billion
Exported Goods and Their Value	Singapore exports petroleum products, rubber, electronics, and computers. Their total value is $61.5 billion.	Hong Kong exports clothing, textiles, toys, watches, electrical appliances, and footwear. Their total value is $145.1 billion.	South Korea exports footwear, clothing, fish, textiles, automobiles, electronics, ships, and steel. Their total value is $80.9 billion.	Taiwan exports textiles, electrical machinery, and plywood. Their total value is $85 billion.

Sources: *1996 Information Please Almanac*, 1995; *National Geographic Picture Atlas of Our World*, 1993

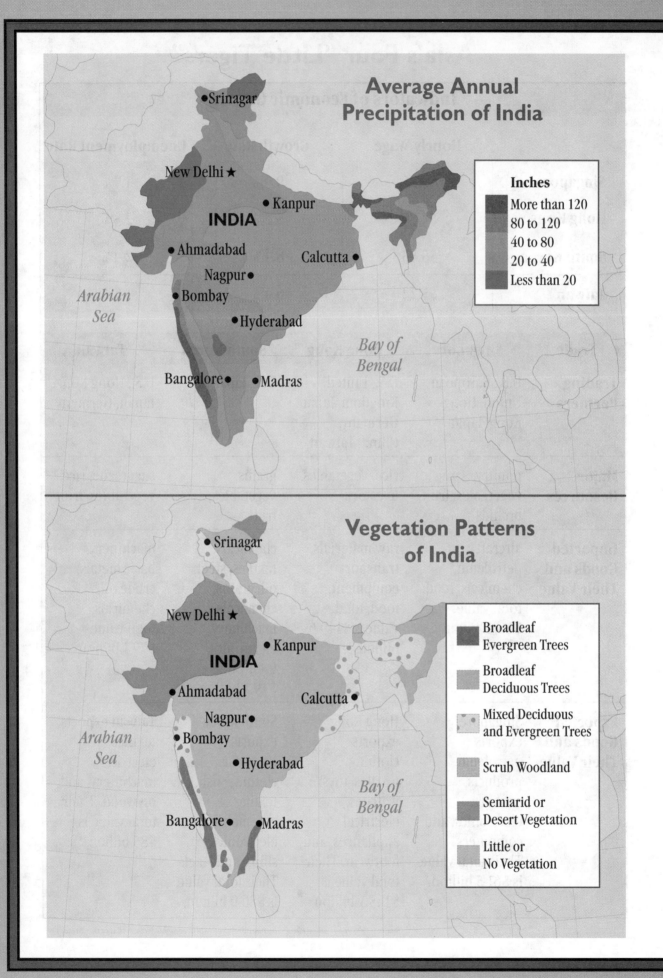

Average Annual Precipitation of India

Srinagar

New Delhi ★

INDIA

Kanpur

Ahmadabad

Calcutta

Nagpur

Bombay

Hyderabad

Arabian Sea

Bay of Bengal

Bangalore

Madras

Inches

More than 120
80 to 120
40 to 80
20 to 40
Less than 20

Vegetation Patterns of India

Srinagar

New Delhi ★

INDIA

Kanpur

Ahmadabad

Calcutta

Nagpur

Bombay

Hyderabad

Arabian Sea

Bay of Bengal

Bangalore

Madras

Broadleaf Evergreen Trees

Broadleaf Deciduous Trees

Mixed Deciduous and Evergreen Trees

Scrub Woodland

Semiarid or Desert Vegetation

Little or No Vegetation

Countries in which Arabic Is the Main Language

1. Algeria
2. Bahrain
3. Chad
4. Comoros
5. Djibouti
6. Egypt
7. Iraq
8. Jordan
9. Kuwait
10. Lebanon
11. Libya
12. Mauritania
13. Morocco
14. Oman
15. Qatar
16. Saudi Arabia
17. Somalia
18. Sudan
19. Syria
20. Tunisia
21. United Arab Emirates
22. Yemen

Countries in which Islam Is the Main Religion

1. Afghanistan
2. Algeria
3. Azerbaijan
4. Bahrain
5. Bangladesh
6. Brunei
7. Chad
8. Comoros
9. Djibouti
10. Egypt
11. Gambia
12. Guinea
13. Guinea-Bissau
14. Indonesia
15. Iran
16. Iraq
17. Jordan
18. Kyrgyzstan
19. Kuwait
20. Kazakhstan
21. Lebanon
22. Libya
23. Malaysia
24. Maldives
25. Mali
26. Mauritania
27. Morocco
28. Niger
29. Nigeria
30. Oman
31. Pakistan
32. Qatar
33. Saudi Arabia
34. Senegal
35. Somalia
36. Sudan
37. Syria
38. Tajikistan
39. Tunisia
40. Turkey
41. Turkmenistan
42. United Arab Emirates
43. Uzbekistan
44. Western Sahara
45. Yemen

Source: *Planet Earth*, Jill Bailey and Catherine Thompson, 1993

127

Spotlight on Milan, Italy

Physical Location	Located in the Alps, in the Po River Valley, near a pass that provides a transportation link between southern and northern Europe.
Population	1,432,200 people in 1991; almost three percent of the total population of Italy. Only the capital city of Rome has more people.
Population Density	

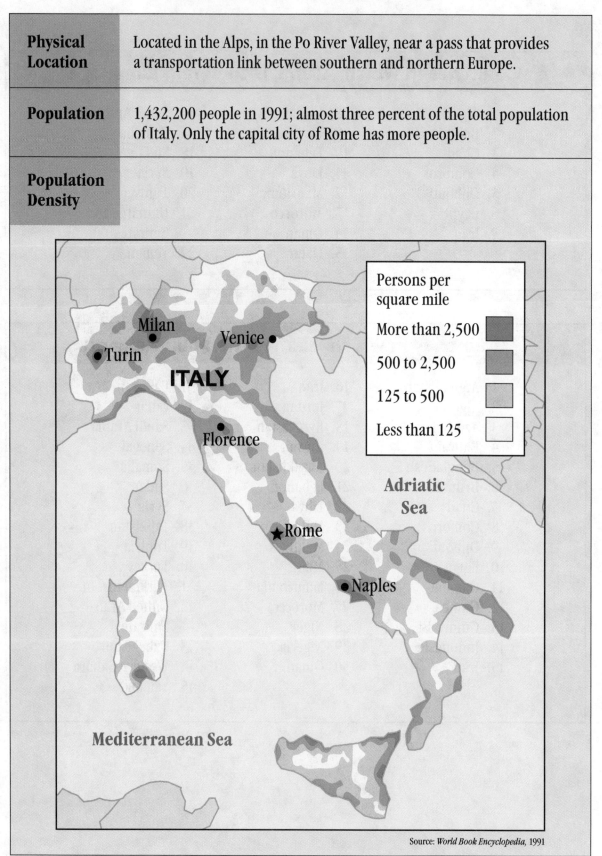

Persons per square mile

More than 2,500

500 to 2,500

125 to 500

Less than 125

ITALY

Milan
Turin
Venice
Florence
Rome
Naples

Adriatic Sea

Mediterranean Sea

Source: *World Book Encyclopedia*, 1991

Spotlight on Milan, Italy

Religion Milan Cathedral	About 95 percent of Italy's population is Roman Catholic. The Milan Cathedral, is one of Europe's largest churches. Many important people in the history of the city are buried in the cathedral. 
Culture	Music lovers from around the world can hear concerts at La Scala, one of the leading opera houses of Europe. The Ambrosian Library contains rare books and ancient manuscripts. Famous Italian paintings hang in the Brera Art Gallery, the Gallery of Modern Art, and the Poldi Pezzoli Museum. Milan has several major universities, including Bocconi University, the University of Milan, and the Catholic University of the Sacred Heart.
Economics	Milan accounts for nearly one third of Italy's total national income. The city is the site of Italy's stock market and the headquarters of major banks. Growing industries include farm equipment, television sets, fine silk, pharmaceuticals, chinaware, and shoes. The Italian economy has surpassed that of the United Kingdom and will most likely surpass other strong European economies, such as France and Germany, in the future.
Politics	The development of Milan in the northern region of Italy has created political problems. Many people have left southern Italy where jobs are scarce, to move north to Milan. This has led to what many people call the two countries of Italy. The richer north has close ties to Western Europe, and the poor south has few growing industries. Many people in the north do not want public funds given to the poor south. This has created a northern movement to separate from the rest of Italy.

European Languages and Religions

Language	The majority of people in present-day Europe speak a language from one of three language subgroups in the Indo-European language family. These language subgroups are Romance, Germanic, and Slavic.

Romance	Germanic	Slavic
Spanish Portuguese French Italian Romanian	German English Dutch Norwegian Swedish	Polish Czech Slovak Ukrainian Russian

Religion	The majority of people in Europe are Christians, usually Roman Catholic, Protestant, or Eastern Orthodox.

Roman Catholic	Protestant	Eastern Orthodox

Spain
Portugal
France
Italy
Belgium
Ireland
 	
large parts of:
   the Netherlands
   Germany
   Austria
   Poland
   Hungary
   Czechoslovakia
	northern Germany
	Great Britain
	the Netherlands
	Norway
	Sweden
	Denmark
	Iceland
	Finland
	Poland
		Greece
		Bulgaria
		Romania
		Serbia
		Ukraine
		Russia

Judaism, a non-Christian religion, is represented by small groups of people throughout Europe. Islam, another non-Christian religion, is represented in small areas in Albania and others parts of the Balkan Peninsula (which includes Croatia and Bosnia).

Asian Languages and Religions

Language	Though English is spoken in parts of Asia, there are many languages that are unrelated to those spoken in Europe. Asian languages are derived from different language families.
	The two main language families are Altaic and Sino-Tibetan. Within each of these language families are several language subgroups. The presence of so many different languages from different families and subgroups makes it difficult to put the languages into broad categories like we did for Europe.
Religion	Christian religions are practiced in some parts of Asia, but the majority of Asians practice non-Christian religions. These include Buddism, Hinduism, Islam, traditional faiths, and tribal religions.

Buddhism	Hinduism	Islam

Myanmar
Thailand
Cambodia
Laos
Tibet
Mongolia
	India
	Nepal
	Bhutan
	Sri Lanka
	Singapore
		Pakistan
		Bangladesh
		Afghanistan
		Saudi Arabia
		Iraq
		Iran
		
		parts of:
		    China
		    the former Soviet Union

Traditional faiths and tribal religions, which are based on ancient, community beliefs, are practiced mostly in northern and eastern Asia, and in parts of Saudia Arabia.

Sources: *Geography: Realms, Religions, and Concepts*, H.J. de Blij and Peter O. Muller, 1994; *The Cambridge Encyclopedia of Languages,* David Crystal, 1994; *Human Geography: Landscapes of Human Activity*, J. Fellmann, A. Getis, and J. Getis, 1992

Profiles of Pacific Islands Countries

Country	Geographic Area	Topography	Chief Crops	Language
Fiji	7,056 sq. mi.	tropical rain forest, coastal plains, valleys	sugar cane, coconuts	English
Kiribati	313 sq. mi.	low-lying, coral and rock	coconuts, breadfruit	English, Gilbertese
Mariana Islands	396 sq. mi.	coral reefs	coconuts	English
Nauru	8.2 sq. mi.	plateau of phosphate	none	Nauruan, English
New Zealand	104,454 sq. mi.	mountainous, fertile plains, volcanic plateau, glaciers	grains, fruits	English, Maori
Papua New Guinea	178, 704 sq. mi.	thickly forested mountains, coastal lowlands	coffee, coconuts, cocoa	English, Melanesian, Papuan
Tuvalu	9.4 sq. mi.	low-lying atolls, none more than 15 ft. above sea level	coconuts	Tuvaluan, English
Western Samoa	1,093 sq. mi.	mountainous	cocoa, coconuts, bananas	Samoan, English

Profiles of Pacific Islands Countries

Country	Population	Ethnic Groups	Settlement History/Government
Fiji	787,000	Melanesians	British, 19th century/Republic
Kiribati	78,000	Miconesians	British, 19th century/Republic
Mariana Islands	176,500	Chammaros (native islanders)	Spanish, 17th century/Germany, United States Commonwealth, Japanese, UnitedStates, 20th century
Nauru	10,000	Nauruans, Pacific Islanders	British, 18th century/ Republic Australian, Japanese, 20th century
New Zealand	3,389,000	Europeans/British Polynesians/Maori	Maoris, 14th century/Parliamentary Dutch, 18th century/Democracy British, 19th century
Papua New Guinea	4,197,000	Papuans, pygmies, Melanesians, Chinese, Australians	Europeans, 5th century/Parliamentary Dutch/Democracy 19th century; Britain, Australia, Germany, 20th century
Tuvalu	10,000	Polynesians	British/Constitutional Monarchy
Western Samoa	204,000	Samoans, Euronesians	Germany, 19th century/ Constitutional Monarchy New Zealand, 20th century

GLOSSARY

A

acid rain: formed by precipitation and the fumes from burning sulfur and nitrogen oxides

archipelago: a group or chain of islands

atoll: a circular or nearly circular coral reef or string of reefs rising above and surrounding a lagoon

B

barchans: horseshoe-shaped sand dunes

C

canal: an inland waterway built to carry water for the purpose of navigation or irrigation

D

desert: a place where more water is lost due to evaporation than falls as precipitation

desertification: the process of becoming a desert through climate change or mismanagement

E

East: a reference to regions having a culture derived from ancient non–Europeans, especially Asian areas

ecology: relationships between the living and nonliving things in an environment

economic growth rate: rate per year at which the economy is growing

ecosystem: a group of living organisms that depend on one another and the environment in which they live

ergs: an Arabic word to describe huge dune fields formed by wind patterns

erosion: the gradual wearing or washing away of the soil and rock of Earth's surface by glaciers, running water, waves, or wind

escarpment: a steep slope or cliff formed by erosion or faulting

fossil water: water that has been underground since before the desert was formed; once the water is gone, it cannot be replaced

free port: a place where goods can be unloaded, stored, or manufactured without import taxes having to be paid

glaciers: large sheets of ice

high islands: islands that are the peaks of sunken underwater mountains; they receive large amounts of rain, are thickly forested, and have active volcanoes

hyperarid: a very dry region that receives less than one inch of rain per year

Ice Age: any prehistoric period when ice sheets and glaciers covered much of Earth's surface; last Ice Age was 18,000 years ago

International Date Line: an imaginary line running approximately along the 180th meridian, marking the time boundary between one day and the next

isthmus: a narrow strip of land bordered by water and connecting two larger bodies of land

lagoon: a shallow area of ocean water partly or completely enclosed within an atoll; shallow body of seawater partly cut off from the sea by a narrow strip of land

leeward side: the side protected from the wind; direction toward which the wind is blowing

linear dunes: dunes that are formed by a combination of winds coming from different directions that often lie parallel to each other, making the landscape look striped

lock: an enclosure, as in a canal, with gates at each end used in raising or lowering boats as they pass from one level to another

low islands: islands made of coral reefs that lie on underwater peaks; they often have poor soil, low rainfall, and lack of fresh water

savanna: a tropical grassland area

silt: a layer of fine mud

special purpose maps: maps which focus on a specific subject

star dune: a large, single dune with ridges forming a peaked summit

strait: a narrow water channel connecting two larger bodies of water

superimposing maps: putting one map on top of another

monsoon: the Arabic word for *season;* a wind that changes direction at regular times of year that blow over northern India and the surrounding area

taiga: an evergreen forest of Siberia

terrain: a region or tract of land, especially with regard to its natural features or suitability for some special purpose, such as farming

topography: physical features

toxic: of, relating to, or caused by poison

tributaries: small water channels that carry water into a river

tsunami: waves in the ocean caused by earthquakes

plateau: an area of relatively flat land elevated above the surrounding land

pollutants: something that pollutes, especially industrial waste or other material that contaminates the air, water, or soil

region: an area of Earth having one or more common characteristics; can be physical or human characteristics

relief: the changes in elevation between parts of Earth's surface

wadis: channels cut through the desert by water

watershed: area drained by a river system

West: a reference to regions having a culture of the Americas or Europe

westernize: to cause to adopt customs, ideas, or qualities regarded as characteristics of Europe and the Americas

windward side: located on or moving toward the side from which the wind is blowing

INDEX

A. Study the climate and vegetation map of the Ice Age and answer the following questions.

1. What was the general moisture of North America like during the Ice Age?

same as today

2. What was the general moisture of present-day United States like during the Ice Age?

wetter than today

3. What was the primary vegetation during the Ice Age?

spruce-rich forest

4. What part of present-day North America was covered with ice sheets?

northeastern United States and Canada

B. Study the present-day climate and vegetation map to answer the following questions.

1. What is the primary vegetation in the United States on the present-day map?

oak-rich forest

2. Where is the spruce-rich forest today?

Canada and Alaska

C. Use both climate and vegetation maps to answer the following questions.

1. How did the ice affect present-day vegetation?

The ice melted, producing water that created spruce-rich forest in Canada and oak-rich
forest in the eastern United States.

2. How would you describe the change in the patterns of forests between the Ice Age period and present-day? Explain the role melted glacial ice played in the change in vegetation.

During the Ice Age, the ice sheet extended as far south as the Midwest, covering the Great
Lakes, with a few spruce-rich forests below that point. In the present day, much of the ice
sheet has melted, uncovering the midwestern United States and Canada, causing spruce-rich
forests to grow across Canada and oak-rich forests to grow in the eastern United States.

Lesson 1

ACTIVITY Describe how the physical environment affects the way people live.

The Impact of the Environment

The **BIG** Geographic Question   How does the environment affect the way people live?

From the article you learned how Ice Age people coped with life in a cold climate. The map skills lesson showed you the climate's effect on vegetation during the Ice Age and the present day. Find out about how the physical environment influences people's lives.

A. Using information from the article, think about characteristics of the Ice Age environment. Complete the following.

1. Describe the land that was not covered by glaciers during the Ice Age and describe where this land was located.

The land south of the glaciers, most of the southern and western parts of North America, was

treeless and dry.

2. What physical conditions did people who lived during the Ice Age encounter and how did they adapt to those conditions?

They encountered frozen terrain and a harsh, cold climate. Vegetation and water were sparse and were

found farther south in North America and west in Asia. The people stayed warm by living in caves,

building fires, and using animal hides for clothing. They also hunted, gathered, and fished for food.

3. What were the movement patterns of people during the Ice Age, and what factors influenced them?

The people of the Ice Age migrated south in the winter and north in the summer. Later they migrated

across a land bridge between North America and Asia that existed at that time. They migrated across

the Bering Strait land bridge when the animals they herded and hunted became scarce in Asia.

B. Using the data from the article and the map skills lesson, write three generalizations about how the physical environment influenced the shelter, clothing, and food of Ice Age people. Then write three generalizations about how the environment affects people today.

	Ice Age	Present Day
Shelter	Ice Age people found caves and built hide tents for shelter.	Today people build houses and apartments out of wood, bricks, and other materials that are now available from the environment or manufacturing.
Clothing	They made warm clothes from the skins of animals that lived during that time.	They wear light fabric clothes in warm weather and layers of clothes in cold weather.
Food	They made weapons and tools for hunting and fishing.	People either buy many types of foods in stores or grow some of their own foods. Foods bought in stores are grown on large farms. When there is a problem, like a flood or drought, these big farms lose crops, so the food prices are higher in stores.

C. From the chart you can see how the physical environment affects the way people live. But the reverse is true, too—people affect the environment in which they live. Which of the two previous statements do you think most appropriately applies to the Ice Age and which to the present? Explain your answer.

The Ice Age was a period of time in which the environment affected the way people lived because

people had to adapt to the physical environment by building shelters, gathering and hunting food, and

making clothing from resources that were found in the frozen environment. Today, people mostly adapt

their environment to meet their needs. For example, people use the resources of the land to build,

invent, or create new materials to meet their basic food, shelter, and clothing needs.

A. Maps show ocean depth and area at sea level. Study the map and legend and complete the following.

1. The cities on the map are at sea level, the place where land is the same height as the sea.

a. List the cities on the map that are at sea level.

Tokyo, Honolulu, San Francisco, San Diego, Los Angeles, Aleutian Islands, and Seattle

b. Locate and circle Japan and Hawaii. (Make sure students accurately circle these islands.)

2. The colors on the color-code strip show the depths of the ocean.

a. Circle the color on the strip that represents 20,000 feet or more. (Make sure students accurately complete the following.)

b. Draw an arrow from the strip you circled to cities on the map that are near that color.

3. Based on the fact that tsunamis originate in the deepest parts of the Pacific Ocean, in what places are earthquakes and volcanic eruptions likely to occur?

Tokyo, Japan; Aleutian Islands, Alaska; and Honolulu, Hawaii

B. More than three fourths of the world's earthquakes occur around the Pacific Ocean. This area around the Pacific Ocean is called the "Ring of Fire" because of its frequent volcanic eruptions.

1. Describe how tsunamis, earthquakes, and volcanic eruptions are connected.

Tsunamis, earthquakes, and volcanic eruptions are all disruptive tremors or movements that occur on

the surface of Earth. They all occur along the "Ring of Fire," along the Pacific Ocean and are very

destructive.

2. Explain why Japan, Alaska, and Hawaii's relative location makes them likely targets for tsunamis.

Japan and Alaska are located near the deepest parts of the Pacific Ocean, within an earthquake zone

and near active volcanoes which create tsunamis. Hawaii is surrounded by these faraway tsunami,

causing earthquake zones and volcanoes.

Lesson 2

ACTIVITY Create a model of a tsunami and compare its effects with those of a hurricane or typhoon.

Creating a Tsunami

The **BIG** Geographic Question   How do the effects of a tsunami compare with those of a hurricane or typhoon?

From the article you learned what a tsunami is and what causes it. The map skills lesson showed you how to identify ocean depths, which affect tsunamis. Now create a model of a tsunami and do research to show how it compares to a hurricane or a typhoon.

A. Like a tsunami, hurricanes and typhoons are natural hazards that form in the oceans. Look in the Almanac to find out more about a hurricane or a typhoon and answer the following questions.

1. How does a hurricane or typhoon form? A hurricane or typhoon forms when an easterly wave of

low pressure deepens and intensifies, causing wind to circulate and move westward, growing stronger

and larger.

2. How big can it get? It can get as big as 200–300 miles across.

3. How does water move as the hurricane or typhoon occurs? Water swirls around the eye of a

hurricane or typhoon.

4. What effects can a hurricane or typhoon have? The strong winds and heavy rains can cause

flooding, land and building destruction, and death.

C. Compare what you have found out about hurricanes or typhoons with what you know about tsunamis, and complete the chart below.

Natural Hazard	Tsunami	Hurricane/Typhoon
Cause	earthquake; volcanic eruption	an area of low pressure deepens and intensifies
Size	wave height varies from a few feet to several hundred feet	at least 74 miles per hour; winds can reach 130–150 miles per hour
Effect in Water	displaces water and causes fast-moving, small waves	water is concentrated in the storm's eye
Effect on Land	can destroy buildings and cause many deaths	flooding and destruction of land and buildings; deaths

D. Assemble the materials you will need to create your three-dimensional model of a tsunami.

1. Look again at the diagram of a tsunami on pages 8–9. Then think about what kinds of materials—cardboard, paper, clay, papier-mâché, styrofoam—would work best to show each stage of a tsunami and its effects.
(Models should show cause, size and shape, and effects of a tsunami.)
2. Display your model in a box, on a flat board, or on cardboard. Make labels that explain what a tsunami is and its different stages.
(Models should include labels for the different stages of a tsunami.)

A. Trace the route of Xenophon from Cunaxa to Byzantium. Use a piece of string to measure on the map the distance between the following places along the route. Then use a ruler to measure the string in inches. Use the map scale to convert inches to miles.

Distance Between	Number of Miles
1. Cunaxa and Nineveh	400 miles
2. Nineveh and Trapezous	600 miles
3. Trapezous and Byzantium	800 miles

B. Use information from the map to answer the following questions.

1. What was the closest city to the battle at Cunaxa? Babylon

2. What river did Xenophon's army follow from Cunaxa to Nineveh? Tigris

3. In what desert did the battle of Cunaxa take place? Desert of Mesopotamia

4. What was the first mountain chain the army crossed on its journey to Cunaxa? Taurus Mountains

5. What body of water did the Greek army sail to reach Heraclea? Black Sea

C. Draw conclusions about Xenophon's route by using information from the map.

1. Note the topography, or physical features, along the route. What was probably the easiest part of the journey to travel on foot?

The middle part of the journey, through the Desert of Mesopotamia, was probably the easiest.

2. Why might the trip north from Nineveh have been particularly difficult?

It may have been particularly difficult because the area is very mountainous.

3. Why would the fact that Asia Minor is located on a peninsula have helped the Greeks in their journey?

A peninsula is surrounded on three sides by water. Asia Minor is surrounded by the Black Sea,

the Aegean, and the Mediterranean. The Greeks were able to sail the Aegean to Asia Minor. They

were able to navigate the Black Sea to reach the Aegean on their way back.

Lesson 3

ACTIVITY   Explain how a military event in history was affected by xenophobia.

The Significance of Xenophobia

The **BIG** Geographic Question   What physical and human elements influenced the outcome of Xenophon's march?

In the article you read about how Xenophon's army faced many problems as it struggled to get back to Greece from Persia. In the map skills lesson you learned about how the land's physical features made it difficult for the army to travel. Now find out how xenophobia could have affected and changed the outcome of historical military events.

A. Answer the following questions using information from the article and the map skills lesson.

1. Define *xenophobia*.

Xenophobia is a fear of strangers.

2. Describe the physical features of the land through which Xenophon's army passed when it encountered the Kurds.

It had to pass through the rugged, unfamiliar terrain of high mountains, hidden valleys, and deep rivers.

3. What mountain group was xenophobic? What happened when this group saw Xenophon's large army?

The Kurds were xenophobic. They took their families into the hills and tried to stop the Greeks from

moving toward Armenia.

4. Why would this mountain group fear the approaching army?

They feared army would kill their families and take their land.

5. How did xenophobia affect the Greeks as they traveled through Persia?

The Greeks needed guidance in traveling across unfamiliar mountains and territory. The Kurds' fear

of Xenophon's army as strangers prevented the Kurds from helping them.

B. Think about the benefits and drawbacks of Xenophon's army and the Kurds fearing each other. Organize your thoughts in the chart below.

Group	Benefits of Xenophobia	Drawbacks of Xenophobia
Xenophon's Army	fear allowed for domination of people; could have lead to surrender of opposing troops	could have been left stranded and vulnerable to attack by native people because they didn't know physical terrain
Kurds	created caution when encountering an approaching army that might attack, which could have lead to survival	prevented the exchange of ideas and trade; isolated one culture from another

C. Write a short skit describing an encounter between the following two historical characters. Answer the following questions to help you plan your skit. Use dialogue to help develop your characters' relationships. Be sure to use historical information from the article.

(Students' skits should reflect perspectives of the historical characters and how they would relate with each other.)

• a Kurd in Turkey facing troops from Xenophon's army
• a Greek mercenary serving in the army with Xenophon facing a group of Kurds

1. How do you feel about meeting these strangers? Do you fear them? Why or why not?
2. What would you say to them?
3. Would you try to help or hurt these strangers? How could you help them?
4. How would your actions be different from or similar to those taken during Xenophon's time?

D. What do you think was the importance of xenophobia in the historical march of Xenophon?

(Answers should include the idea that because many native peoples were xenophobic, physical

features were not easily overcome during the march.)

A. Study the two maps. Circle the following places on both maps.
1. Rome
2. Carthage
3. Italy
4. Tunisia

B. Study both maps to answer the following questions.

1. In which modern country is the city of Rome located? Italy

2. Locate the ancient city of Carthage. In which modern country was it located? Tunisia

3. What present-day city appears to be located in the same place as ancient Carthage?

Tunis

4. Into which modern European country did Carthage's power extend before the Punic Wars? Spain

5. Carthage colonized along the coastlines of which three northern African countries?

Tunisia   Algeria   Morocco

6. What two modern countries does the Strait of Gibraltar separate?

Morocco   Spain

C. Compare the information on the two maps to draw conclusions about how they differ.

1. How does the area of Roman domination compare on the two maps?

Rome dominated the entire Italian peninsula in ancient times, but is one of many cities in Italy today.

2. How does the area of Carthaginian domination compare on the two maps?

Carthage dominated the northern coast of Africa and much of modern Spain. Carthage does not exist

on the map today.

3. What boundaries exist on the modern map that do not exist on the ancient map?

national boundaries of European and African countries

4. Given what you learned in the article, why do you think Carthage does not appear on the second map?

Rome permanently destroyed Carthage.

Lesson 4

ACTIVITY   Create a resource map of the Carthaginian empire.

When Carthage Was Wealthy

Geographic Question

You learned from the article about the wealth and power of the ancient city of Carthage. The map skills lesson showed the area Carthage ruled. Find out how the city's location and its access to natural resources contributed to its wealth and power.

A. Use information from the article and map skills lesson to answer the following questions.

1. Which modern countries occupy the area that Carthage covered?

Tunisia, Algeria, Morocco, Spain, and Italy

2. What are the major physical features of the area surrounding the ancient city of Carthage?

Carthage was located on the southwestern European and northern African coasts. In the middle

of its land was the Mediterranean Sea and the Strait of Gibraltar. The Atlas Mountains to the south

separated the city from the Sahara Desert.

3. How did these major physical features help Carthage maintain its position of wealth and power?

The Carthaginians were able to use the Mediterranean Sea and Strait of Gibraltar to explore and

colonize the west coast of Africa, islands of the Mediterranean, and part of present-day Spain,

thus expanding their empire and trade markets. The Atlas Mountains to Carthage's south protected

them from being invaded and taken over from that side.

B. Using the Almanac find out about the resources of the Carthaginian Empire. In the space below draw a symbol to represent each resource.

(Make sure students' symbols help identify the resource they represent.)

C. Use the information you've collected on physical features and resources to fill in the map of the Carthaginian Empire in the Almanac. Put the symbols you created in the map key. You may wish to color-code the physical features and provide a key to each color's meaning. Don't forget to give your map an appropriate title. (Students' maps should accurately include resources, symbols, and physical features of ancient Carthage.)

D. Explain what your map shows about how Carthage became powerful and wealthy!

(Students' answers should include the fact that controlled land on two sides of the Mediterranean Sea

gave Carthage lots of power to use the sea for travel, trade, and defense. Also, Carthage's control of the

Strait of Gibraltar, which gave it access to the Atlantic Ocean, made the country powerful and wealthy.)

LESSON 5

A. Look at the map of Europe. Circle the following cities and study their locations.
1. Zurich, Switzerland   4. Copenhagen, Denmark
2. Munich, Germany   5. Dresden, Germany
3. Oslo, Norway

B. Answer the following questions about the cities listed above.

1. Which of the five cities are landlocked? Dresden, Munich, and Zurich

2. What large bodies of water are the remaining two cities located near? Oslo is near the North
Sea, while Copenhagen is on the Baltic Sea.

3. Between the two cities located near large bodies of water, which would you consider to be the
most favorably located? Oslo

4. Why is the location of this city a favorable location? Oslo is located on an arm of the North Sea
which flows into the Atlantic Ocean. Its location makes it easier for people and goods from different
countries to get in and out of the city.

C. Analyze and indicate each city's location on the chart below.
Then indicate on the chart the advantages and disadvantages
of cities being located close to or far from these features.

Cities	Location	Advantages	Disadvantages
Zurich, Switzerland	central Europe near the Rhine River; in the Alps	mountains offer protection from invasion; center of Europe, crossroads of trade	difficult to travel through mountains; landlocked
Munich, Germany	southeastern Germany near the Danube River; in the foothills of the Alps	Danube River provides transportation link for trade and travel	foothills don't provide complete protection from invasion; landlocked
Oslo, Norway	southeastern Norway along the coast of the North Sea; no mountains	access to sea routes to Europe and the world through the North Sea	more open to invasion since it's accessible by sea and ocean
Copenhagen, Denmark	eastern coast of Denmark, at the entrance of the Baltic Sea; no mountains	controls entrance into Baltic Sea; ideal defensive position	being surrounded by water makes it hard for a city to expand; little unused land
Dresden, Germany	eastern Germany on the Elbe River; no mountains	the Elbe provides transportation links for trade and travel	difficult to defend because of its position with no natural barriers around it

LESSON 5

Lesson 5

ACTIVITY Identify the reasons for the growth of a city in Europe today.

A City on the Rise

The **BIG** Geographic Question   What factors might influence the growth of a city in Europe today?

From the article you learned why and how cities developed in Europe in the past. The map skills lesson helped you locate cities of Europe in relation to nearby physical features. Now determine why and how Milan, Italy, has become one of the fastest growing cities in Europe today.

A. Use what you have learned about how cities developed in the past to answer the following questions.

1. Name four characteristics that influence the function of a city.

location _____ accessibility _____

resources _____ inhabitants _____

2. Name two reasons why rivers were important to the location of cities.

A location on a river provided an easily defensible position and transportation for trade.

_____

_____

3. How did population growth help the development of capital cities in Europe?
Large populations required local government. Cities that became centers of government
became capital cities as Europe was organized into countries.

_____

_____

LESSON 5

4. Why were hospitals and universities established as European cities grew?
Hospitals and universities were established to provide medical care and education to
the growing population.

B. Using the information from the article, the map skills, and the Almanac,
complete the following chart about Milan.

Physical Features	located in the Alps; located in the Po River Valley; near a pass that provides a transportation link between northern and southern Europe; Po River
Economic Features	accounts for one third of Italy's total income; site of Italy's stock market; headquarters of major banks; growing industries include farm equipment, TVs, silk, shoes, chinaware, and pharmaceuticals
Political Features	close ties to Europe; destination for many poor, unemployed people in southern Italy; northern movement to separate from the south because the north does not like public funds being given to the south
Cultural Features	one of Europe's largest churches, the Milan Cathedral, located here; many important people in the history of the city are buried in the Cathedral; La Scala, one of Europe's leading opera houses; Ambrosia Library houses rare books; several major universities
Population Features	1,432,200 people in 1991; almost 3 percent of the total population of Italy

C. Use the information you have gathered to write a paragraph about why
you think Milan is rapidly growing as a leading European city. Think about
its location and how it affects the activities within the city.

(Students' answers should reflect an understanding that the physical location of Milan has been the key
to its development. Physical location includes the city's proximity to Western European markets.)

_____

_____

_____

LESSON 6

A. Look at the map of Europe. Circle the following on the map.
1. Rhine River
2. France
3. Germany
4. The Alps (mountains)
5. North Sea
6. Bonn
7. Moselle River

B. Answer the following questions about the rivers of Europe.

1. What mountain range is the source of the Rhine River? the Alps

2. Which countries does the Rhine River flow through or border? Switzerland, France,
Germany, and the Netherlands

3. What major cities are located on the Rhine River? Rotterdam, Duisburg, Cologne, Bonn,
and Strasbourg

4. Where is the mouth of the Rhine River? in the North Sea

5. What smaller rivers flow into the Rhine River? Lippe, Ruhr, Main, Moselle, and Neckar

6. About how long is the Rhine River? about 800 miles

7. In what direction does the Rhine River flow? mostly north

8. Where is the Rhine River in relation to the Rhône River? (north, south, east,
or west) north and east

9. Why do you think the Rhine River is so important to Europe? The Rhine flows through many
countries. Several important cities are built on this river. It helps make Europe's interior easier
to reach for transportation and trade.

_____

LESSON 6

Lesson 6

ACTIVITY Rate Europe's rivers.

A Rave River Review

The **BIG** Geographic Question   What are the most important features of rivers?

From the article you learned about the rivers of Europe. The map skills lesson showed you the course of the Rhine River and its importance to Europe. Which of Europe's rivers best serves the continent? What characteristics of each river are important? (Students' answers should indicate an understanding of how a river flows and the effect it can have on a region.)

A. Look at the physical map of Europe and the information about
its rivers shown in the Almanac. Then take notes on the following.

Source
1. As is true for most rivers, what is likely the source for Europe's many rivers? The likely source
for Europe's rivers is a place of high elevation, or in the mountains.

Mouth
2. Where are the most useful locations for the mouths of Europe's rivers? Useful river mouths are
those with access to oceans or seas so that goods can be shipped more easily and widely.

Trade
3. What is the best route for a European river for it to be valuable for trade? The best river trade
routes are those that pass through areas with large populations and have access to other water
routes such as oceans or seas.

Population
4. What is the best route for a European river to follow for it to serve the
most people? Rivers that flow through populated areas and areas with resources help transport
goods and facilitate trade.

Tributary
5. What are the best routes for a European river's tributaries for them to have
an important effect on the area through which they flow? The best routes for tributaries are
those that flow through areas where people live that are not near major rivers.

LESSON 6

B. After reviewing the information about Europe's rivers, complete the
chart below to determine the positive and negative features of each river.

River	Positive Features	Negative Features
Volga	longest of Europe's rivers; many cities located along its banks; most important internal waterway in Russia; has almost 200 tributaries	flows to Caspian Sea, but the sea is landlocked (no access to the open oceans)
Danube	flows through 9 countries; banks serve as the site of some of Europe's greatest cities; has almost 300 tributaries	flows mostly through farm areas instead of areas of heavy industry (areas of heavy river traffic)
Dnieper	flows through a region rich in natural resources and a major industrial region in Ukraine	only flows through Belarus and Ukraine; does not link these two countries with Western Europe
Rhine	widely used by large pop. within its basin; flows through industrial regions; linked to world commerce through Rotterdam; receives water from several smaller rivers	shorter than Danube River; flows through fewer countries than the Danube
Elbe	provides a connection for the Czech Republic to Germany and the rest of Europe	short river; flows only through the Czech Republic and Germany
Rhône	only European river that flows directly to the Mediterranean Sea; provides a major connection between Europe and the Mediterranean	short river; flows in a difficult, zig-zag course

C. Now use the information you have gathered to "grade" each river and
determine which is the most valuable to Europe. Compare your "grade"
for each river with a classmate's. Did your classmate choose the same
or a different river as Europe's most valuable river? Are your "grades"
different? Discuss the reasons for the similarities and differences.

1. Volga _____   4. Rhine _____
2. Danube _____   5. Elbe _____
3. Dnieper _____   6. Rhône _____

(Students' grades should indicate that they considered the number of people and countries served by the rivers, the importance of commerce, and the level of development of the regions through which the rivers flow.)

A. Study the elevation key on the map of Europe and Asia. Use the key and the map to help you identify the following features.

1. How did you use the map to help you identify:
   a. mountain ranges?
   looked for the color shadings that represented areas of high elevation

   b. plains?
   looked for the color shadings that represented areas of low elevation

2. Define and distinguish between a peninsula and an island.
   Peninsulas are extensions connected to the land and are surrounded by water on three sides. Islands
   are surrounded by water on all sides.

B. Look at the map and complete the following.

1. Draw a line showing where Europe and Asia are divided.

2. Using the map, write a description of the physical features you see in Europe.
   Europe is the western portion of the Eurasia landmass. It has some high mountains and is
   surrounded by water on three sides. It has many small peninsulas along its coastline and plains
   on its interior.

3. Why do you think Europe is considered a peninsula of Asia, rather than Asia being considered a peninsula of Europe?
   Asia is a larger piece of land, and Europe appears to jut out from Asia.

4. Write a description of the physical features of Asia.
   Asia is a huge landmass with some high mountains, uplands, and highlands. However, large areas of
   Asia's land are plains and plateaus. Its eastern coastline is bordered by many islands.

5. Describe the physical features that are shared by both Europe and Asia.
   Europe and Asia share the Ural Mountains and a huge plains area. The North European Plain on
   the Europe side of the Urals becomes the West Siberian Plain on the Asia side of the Urals. The two
   continents also share the Caspian and Black seas as partial borders between them.

Lesson 7

ACTIVITY
Organize information to use in a debate, discussion, or report.

Continent Combo?

The BIG Geographic Question
Should Europe and Asia be identified as a single continent or two separate continents?

From the article you learned about the diverse physical features of Europe and Asia. In the map skills lesson you looked at some of these physical features on a map. Now decide whether Europe and Asia should be two continents or one.

A. Use the article, the map skills lesson, and your own ideas to answer the following questions.

1. **Geography:** What physical features connect Europe and Asia? What features separate them? Keep in mind the following physical features: plains, mountain ranges, depressions, rivers, and climate zones. (Possible answers include the following.)
   Connects Europe and Asia: North European and West Siberian Plains; climate (warm summers and
   cold winters in some parts and subarctic and polar in other parts)
   Separates Europe and Asia: Ural Mountains form the boundary between Europe and Asia; Caspian
   Sea lies between Europe and Asia

2. **History:** Why have Europe and Asia been considered separate continents?
   European mapmakers in the sixteenth century established them as a separate continents.

3. **Culture:** What complications might result if Europe and Asia became one continent?
   Maps and history books would be outdated. Communication problems could occur between
   the people of the two continents because of the differences in language, religion, government, and
   currency.

B. Pretend you are a member of a World Mapmakers Association. You are traveling as a delegate to the United Nations to attend a special meeting of their General Assembly. The delegates will present arguments for and against the following proposal. Use the chart below to organize the pros and cons of the proposal. (Possible answers include the following.)

The United Nations should recognize Europe and Asia as a single continent, hereafter to be identified on all maps as Eurasia.

	Geography	History	Culture
Pros	Eurasia could be studied from a geographic point of view because of features common to both Europe and Asia. Maps should reflect the physical geography of the land rather than human history and culture.	Events that occurred in both Europe and Asia could be treated more fully and their connections better understood.	People from the two continents interact and could be studied together. It might open up more possibilities for trade cooperation (such as the European Union with one currency, one language, and so on).
Cons	Current maps and history books would have to be outdated.	History books would have to be updated. Historical animosities over wars and rulers might resurface and be hard to overcome.	The cultures of Europe and Asia, today as in the past, are quite diverse and reflect the traditional division into separate continents.

C. Choose a side on the proposal and organize your arguments in a debate, a discussion, or a short research report.

(Students' chosen projects should take a stand for or against combining Europe and Asia to form

Eurasia. Students should use information concerning geography, history, and culture in their

arguments as support for their stand on the proposal.)

A. Study the maps and scales to answer the following questions.

1. How long is Lake Superior? about 400 miles long

2. How long is Lake Baikal? about 400 miles long

B. Using the maps and the table below to complete the following.

	Lake Baikal	Lake Superior
Area	12,162 sq. mi.	31,700 sq. mi.
Volume	5,581 cubic mi.	2,916 cubic mi.
Elevation	1,493 ft. above sea level	600 ft. above sea level
Deepest point	5,315 ft.	1,330 ft.
Percentage of Earth's fresh water	20%	10%
Economic uses	harbors, shipping route for minerals, resort	commercial fish, forest and mineral resources, resort

1. Judging from the area and depth of each, which lake appears to be larger? Which lake is actually larger? Explain your reasoning.
   Lake Superior appears to be larger because it has a greater surface area; Lake Baikal appears to be
   larger because it is deeper.

2. What is the volume of water for each lake? Which lake do you think has a greater percentage of the world's fresh water? Explain.
   Baikal: 5,581 cubic miles; Superior: 2,916 cubic miles; Lake Baikal probably has a greater
   percentage of fresh water because it contains a larger volume of water.

3. Which lake might serve better as a shipping route? Explain.
   Lake Superior: because it is close to other large lakes, which are connected to major rivers. Also, it
   borders two countries which makes trade between them easier.

4. Write summary statements explaining the similarities and differences between the two bodies of water.
   Both lakes supply fresh water, have a variety of economic uses, are above sea level, and have resort
   areas. Differences include the size, area, depth, volume, and the type of economic activities.

Lesson 8

ACTIVITY
Identify the cause-and-effect relationships associated with the decline of Lake Baikal.

Lake Effects

The BIG Geographic Question
How have human activities caused environmental changes at Lake Baikal?

From the article you read about some of the changes in the physical geography of the Lake Baikal area. In the map skills lesson you compared Lake Baikal to another fresh water lake. Now find out how human activity has turned Lake Baikal into an environmental battleground.

A. Find the following terms in the article on Lake Baikal. Write a definition for each term. Then use the term to write your own sentence about Lake Baikal.

1. ecosystem all living things in one place interacting with their nonliving environment
   The balance of Lake Baikal's ecosystem was upset by overdevelopment.

2. ecology study of the relationships between living things and their environment
   The ecology of Lake Baikal and its surroundings was affected by human actions.

3. erosion wearing away of soil by wind and water
   Soil erosion was caused by the clearing of timberlands surrounding Lake Baikal.

4. toxic poisonous
   The lake's water was contaminated by toxic chemicals.

5. acid rain rain caused by waste fumes and precipitation
   Acid rain contributed to the pollution of Lake Baikal.

B. Develop a picture of Lake Baikal by completing the following chart of what it was like before and after the arrival of the Trans-Siberian Railway.

Before	After
• many species of wildlife • "The Pearl of Siberia" • evergreen forests surrounded the lake • no toxic pollution • beautiful, scenic area • clean natural resource	• serious declines in the size and populations of many species of wildlife • "environmental battleground" • erosion • deforestation • polluted by toxic chemicals • acid rain

C. For each change that occurred after the Trans-Siberian Railway came to Lake Baikal, there was an action that caused it. Below, explore the cause-and-effect relationships and make suggestions for how to prevent the problem from continuing in the future.

Cause	Effect	Possible Solutions
• Trans-Siberian Railway; timber removal and agricultural activity	• deforestation and erosion	• forbid timber removal close to the lake's shore; limit amount of land used for agriculture
• Baikalsk Pulp and Paper Combine	• pollution by toxic chemicals; decline in wildlife population; acid rain	• replant trees as old ones are removed; monitor disposal of toxic chemicals
• factories in Lake Baikal region	• pollution from factory wastes and sewage	• monitor disposal of sewage and other factory wastes; limit the burning of sulfur and nitrogen oxides that cause acid rain

LESSON 9

A. **Label Map A of northern Africa and the Middle East to show those countries in which Arabic is the main language.** (Make sure students follow the directions below.)

1. Look in the Almanac to find countries in which Arabic is the main language.
2. Using the world map in the Almanac, locate these countries on Map A and label them using a colored marker or pencil.
3. In the map key on Map A, draw and color in a small square using the marker or pencil. Label this square "Arabic-Speaking Countries."

B. **Look at Map B to see the areas of northern Africa and the Middle East where Islam is the main religion.** (Make sure students follow the directions below.)

1. Use a different color marker or pencil to outline onto Map A the shaded areas on Map B. Then use the marker or pencil to draw a short line in the map key to represent the outline color. Label this line "Countries of Islamic Religion."
2. Note that you have just indicated on Map A the areas of northern Africa and the Middle East where Islam is the main religion.
3. Note that some countries in the area outlined on Map A are not labeled.
4. Use another different color marker or pencil and the world map in the Almanac to label the remaining countries in the outlined area on Map A.

C. **Use your outline map to answer the following questions.**

1. Which covers a larger land area, the Islamic religion or the Arabic language?

   Islamic religion

2. Which countries included on Map A practice Islam but do not speak Arabic?

   Senegal, Guinea, Mali, Niger, Nigeria, Iran, and Turkey

3. What conclusions can you make about those who live in countries in which Islam is the main religion, but who do not speak the Arabic language?

   One can practice Islam without being able to speak Arabic. One can speak Arabic without practicing the Islam religion or being Muslim. The religion and language are independent of each other.

LESSON 9

Lesson 9

ACTIVITY
Develop research to present information on the spread of Islam.

Islam—A Television Special

The BIG Geographic Question
How does a religion spread from one country to another?

In the article you learned about the Arabic language and its role in the growth of the religion of Islam. The map skills lesson helped you see which countries in northern Africa and the Middle East speak Arabic and practice Islam. Now create an informative television special about the Islamic religion.

A. **Using the article and information from the Almanac, gather information for a television documentary about the spread of the Islamic religion. As you read, answer these questions.**

1. In what city did Islam originate? Mecca

2. In what country is that city located today? Saudi Arabia

3. Why is that city the most important location in the Islamic world? Mecca is the place where Muslims believe Mohammed received his visions and messages from God.

4. Who founded Islam? Mohammed

5. What are the five basic practices or duties this founder taught? belief in Allah and Mohammed's role as his prophet; prayers said in the direction of Mecca; charity toward the poor in the form of a tax; fasting; pilgrimage to Mecca

6. In what book can those practices be found? Koran

7. When and how did Islam reach areas beyond the Arab world (e.g., Pakistan, Indonesia, China, Malaysia, the United States)? Islam spread to areas outside the Arab world through the migration of those who possessed the ideas and through trade. Islam moved outside the Arab world in the 1500s and 1600s.

LESSON 9

B. **Make notes on some of the information you feel would be most important to communicate about Islam.** (Possible answers include the following.)

who founded Islam and its history; the location and importance of Mecca; the five basic practices of Islam; the importance of the Arabic language and the Koran to Islam

C. **Remember that television is a visual means of communication. Decide what kind of visual aids (map, picture, diagram, and so on) you will use to illustrate each point in the documentary. Note your type of visual aids here.**

(Students might indicate a map, picture, diagram, book, video, poster, and so on.)

D. **Draw a "storyboard" for your documentary. You can do this by using index cards to sketch one scene at a time. Decide what the viewer will see as each topic is discussed. Write a few lines of dialogue below each sketch. When you have finished the cards, arrange them on a piece of posterboard. Place the cards in the same order that you would present the information in your television special. Use the space below to plan what will on each panel of your storyboard.** (Students' storyboards should highlight important features of Islam.)

LESSON 10

A. **On the map locate the waterways listed below. Draw a symbol to represent a strait next to each one.**

1. Bosporus Strait
2. Dardanelles Strait
3. Strait of Gibraltar
4. Strait of Hormuz
5. Suez Canal

B. **Use the article, the map on page 58, and the Almanac to answer the following questions.**

1. Would shipping be easier to block in the Gibraltar or Bosporus strait? Why?

   The Bosporus Strait would be easier to block because it is more narrow. It is only one half mile wide.

2. What countries rely on the Strait of Hormuz to ship their oil? Iran, Iraq, Kuwait, Saudi Arabia, Qatar, the United Arab Emirates, and Bahrain

3. Why would Russia want control of the Bosporus Strait? The Bosporus Strait gives Russia access to the Mediterranean Sea, and from there to the Atlantic Ocean and the Suez Canal. This would make it easier for Russia to ship goods to other countries of the world.

4. How is the Suez Canal different from the straits in this area of the world? It was not a natural strait but a waterway built by man. It serves the same function as a strait.

5. Why is the Suez Canal important to Israel? It gives Israel easier and quicker access to the Red Sea, Indian Ocean, and countries in eastern Asia.

6. How was the Strait of Hormuz important to the countries involved in the Persian Gulf War? The Strait of Hormuz allowed United States military troops to enter the Persian Gulf, and eventually Saudi Arabia, during the Gulf War. Coming into Saudi Arabia from the coast of the Persian Gulf was easier for the troops because they didn't have to travel through the desert in southern Saudi Arabia.

7. Why has the Dardanelles Strait been important throughout history? The Dardanelles Strait has been important for defense and military campaigns because it provides a short passage between Europe and southwest Asia.

LESSON 10

Lesson 10

ACTIVITY
Describe in detail the region's straits.

A Strait Visit

The BIG Geographic Question
In what ways have straits been important to surrounding countries?

From the article you learned why straits and a canal that serves the same function as a strait are important. The map skills lesson helped you locate several of these waterways and decide how they affect countries around them. Now conduct an in-depth study of one strait.

A. **Select one of the major straits in northern Africa or southwest Asia. Use the article, the maps skills lesson, and the Almanac to collect information about that strait. Concentrate and take notes on the following points.** (Students' answers should reflect some of the information found on page 122 of the Almanac for the straits they selected.)

1. Where is the strait located? What bodies of water does it connect?

2. What role has the strait played in the history of the region?

3. Why is the strait important to the politics, economy, and defense of the area?

4. What route would you take to visit the strait?

LESSON 10

B. **Imagine you are writing an entry about your chosen strait for a tourist guidebook. Sketch a map of your strait. Then use the space below to write a description of it and a good route to travel there. Try to make your description informative and interesting.**
(Students' maps and descriptions should accurately depict the strait they selected for the guidebook entry.)

Description:

LESSON 11

A. Study the map. Using different colored pencils for each item listed below, circle the following on the map. (Make sure students locate and circle the following items on the map.)
1. Windhoek, Namibia and Gaborone, Botswana
2. Kalahari Desert
3. Towns or settlements along the northern route between Windhoek and Gaborone

B. Use the map, scale, and legend to complete the following.
1. If you were flying from Windhoek directly to Gaborone, how many miles would you travel?

about 550 miles

2. What two other methods of transportation could be used to get from Windhoek to Gaborone? In different colors, trace these two routes on your map.

railroad and roads

3. What form of transportation and route would be the quickest and easiest to travel from Windhoek to Gaborone and how long many miles would the route be?

The road route by car from Windhoek, through the towns of Gobabis, Maun, Francistown, and

Mahalapye would be the quickest and easiest. This route would be about 1,080 miles.

4. Describe the physical features you would encounter as you travel from east to west.

a desert, lake, and swamp

5. What supplies would you need to take with you on your trip using the quickest and easiest route? Explain your answer.

(Students should indicate that they need to take extra water and food, light-weight clothes, and

protection from the sun because they have to travel through the Kalahari Desert.)

6. What form of transportation and route would be the longest and hardest to to travel from Windhoek to Gaborone and how many miles would the route be?

The train route southward through the mountains would be the longest and hardest. This route would

be about 1,180 miles. The train would probably travel slower than a car, and traveling through the

mountains would make it even slower. Plus, there are no places to stop along the way.

LESSON 11

Lesson 11
ACTIVITY Show physical and human characteristics of deserts.

Picturing the Desert

The BIG Geographic Question — How are deserts formed, and how do they change over time?

From the article you learned how deserts form and change. The map skills lesson helped you look at a route traveling through Africa's Kalahari Desert. Now make a visual display of a desert.

A. Use information in the Almanac to make two bar graphs showing the precipitation in the Kalahari and Sahara deserts in Africa over the year by completing the following steps. (Make sure students accurately complete the graphs below.)
1. Place the months of the year on the x-axis.
2. Place the number of inches of precipitation on the y-axis.
3. Appropriately label the axes "Months" and "Inches of Precipitation."
4. Give each graph a title.

LESSON 11

B. Compare the bar graphs you have created by answering the following questions.
1. Which desert has more rainfall per year? About how much more?

The Kalahari Desert has more rainfall. It has about 17 inches more rainfall per year than the

Sahara Desert.

2. What are some of the natural causes that make the Kalahari a desert? How are these different from the natural causes that make the Sahara a desert region?

Sandy soils and lack of surface water (lakes, streams, and rivers) make the Kalahari a desert. The

Kalahari lacks adequate soils even though it receives more rain. The Sahara lacks adequate rainfall.

C. Using information in the article and the Almanac, select one of the following ideas and create a poster. (Make sure students select one of the posters suggested below.)
___ 1. Prepare a poster that shows how nature creates deserts. (Show the effects of physical location, drought, and destruction of plant life.)
___ 2. Prepare a poster that shows how people create deserts. (Show how poor farming practices, overgrazing of cattle, and the cutting of trees contribute to decreasing ground cover and increasing desert.)

D. Each poster should have diagrams, pictures, and a few words to explain the main ideas. Plan your poster by taking notes on the following features. (Make sure students' notes and plans for their posters accurately reflect their selections indicated above.)

Diagram Ideas	Picture Ideas	Brief Written Description

LESSON 12

A. Using the map legend, make a list of the resources found in the area of Kenya near Lake Victoria. Remember that resources can include crops, animals, minerals, and natural features that might promote tourism.

cattle	diamonds
coffee	Lake Victoria
copper	national parks
cotton	sugarcane

B. Think about the resources found in the Lake Victoria area and answer the following questions.
1. The land used for farming around Lake Victoria is very dry. When there is no rain, crops such as corn will not grow, and the people have no food. What available resource could farmers use to water their corn in times of low rainfall? How might they transport the resource?

People could build irrigation systems using the water from the lake.

2. Nile perch were added to Lake Victoria so the area would have a product that could be exported. Describe alternative ways that people might have improved the economy of the area.

(Possible answers include: With irrigation the people could have increased the yield of crops that

already grew well, such as tea, coffee, and sugarcane; Lake Victoria could have been developed into a

tourist attraction with beaches, fishing trips, resorts, treks to see exotic animals, or hikes to nearby

mountains and national parks.)

3. List advantages and disadvantages for each of your answers to question 2.

(Possible answers include the following.

Irrigating the land: Advantage—This would provide plenty of water to grow lots of native crops.

Disadvantage—It would cost a lot of money to build the system.

Make Lake Victoria a tourist attraction: Advantage—This would bring people and money into an area

and provide an opportunity to share information about the importance of preserving the area.

Disadvantage—It would cost a lot to build hotels and fix up beaches, and it might change the

character of the area.)

LESSON 12

Lesson 12
ACTIVITY Develop a representation of a food chain or web.

Creating a Food Chain or Web

The BIG Geographic Question — How do the life forms in an area depend upon one another?

In the article you learned how the introduction of Nile perch into Lake Victoria affected the ecology and economy of the area. The maps skills lesson helped you look at other resources of the Lake Victoria area. Now create your own model or representation of the food chain or web of the Lake Victoria area.

A. Before creating your food chain or web, it will be helpful to consider the following questions. Use information from the article and the Almanac to help you answer them.
1. What organisms will be included in your food chain or web of Lake Victoria?

(Students should include the organisms listed on page 121 of the Almanac.)

2. How might the removal of one plant or animal from your chain or web affect the feeding relationships?

Removing one plant or animal forces animals above that level to feed on other available plants or

animals. This shifts the balance of feeding relationships and can cause a chain of feeding problems.

3. How might the addition of one plant or animal to the chain affect the rest of the chain or web?

If the plant or animal is low on the food chain, it gives higher animals more choices of food. If it is

high on the chain, it can cause the extinction or depletion of one or more of the lower species.

4. Suppose environmentalists were writing a book about what has happened at Lake Victoria as a result of the addition of the Nile perch. What title would you suggest for such a book?

(Students' titles should communicate the importance and effects of the Nile perch introduction.)

LESSON 12

5. What lessons can the rest of the world learn from what has happened at Lake Victoria?

The Lake Victoria problem can help people around the world understand the importance of food

chains and webs and understand that altering these systems in any way can have far-reaching and

rippling consequences.

6. Research another environmental situation in which a plant or animal was intentionally introduced or unintentionally brought to an area. What kind of effect did it have on the environment? Was it positive or negative?

(Students might research the zebra mussels that were brought to the Great Lakes area via a ship

from Russia in 1980. The North American mussel's food was eaten by the Russian zebra mussels.

Or students might research the shaggy nelalukas tree seeds that John Gifford brought to the Florida

Everglades from Australia in 1906. The nelalukas grew out of control, forcing out birds and animals

and keeping other plants from growing there.)

B. Using art supplies prepare a model of the food chain or web for the Lake Victoria region. (Students' food chains or webs should accurately show the feeding relationships in the Lake Victoria area.)
1. Use drawing or construction paper or other art supplies to make three-dimensional representations of the creatures that feed on other animal or plant life. Be sure to include cichlids, Nile perch, insect larvae, brine shrimp, and crocodiles. And don't forget humans! Note that each animal may feed on several of the other smaller species.
2. Arrange the plants and animals that you made on a posterboard. Glue them in place.
3. Label each part of the food chain and give it a title.

C. Imagine one of the elements has been removed from your food chain or web. Below, plan a diagram to show what would happen as a result of its removal.

(Make sure students' diagram plans address the removal of one food chain or web element.)

A. Use information from the map of Japan to answer the following questions.

1. What is the current capital of Japan? Tokyo

2. What are its latitude and longitude coordinates? about 36° North, 140° East

3. What are the four main islands of Japan? Hokkaido, Honshu, Kyushu, Shikoku

4. What bodies of water separate Japan from mainland Asia? the Sea of Japan and the East China Sea

5. What body of water separates Japan from North America? the Pacific Ocean

6. Where are the Japanese Alps located? The Japanese Alps are located on the island of Honshu in the central and western sections of the island.

B. Use the map to help you answer the following questions.

1. How many ports did Japan have in 1700? Name them. 2; Edo/Tokyo and Nagasaki

2. How many ports did it have in 1858? Name them. 5; Nagasaki, Hyogo, Shimodo, Edo, and Hakodate

3. How many ports were there in Japan by 1990, and on what side of the islands are the majority of them located? Why are they located there? By 1990 there were 12 ports, and the majority of them are located on the Pacific Ocean, or eastern, side of the islands. The land on the eastern side is not mountainous and rocky like the central and western sections.

C. Using the information you have learned about Japan, explain how it was able to isolate itself from the rest of the world during the 1600s.

As an island nation, Japan is physically separated by water from the rest of the world. It does not share its borders with any other countries. Its terrain is mountainous. These physical features helped the shoguns close the country to outside influence.

Lesson 13

ACTIVITY
Find out how Japan went from being an isolated country to a modern global power.

Advising the Shogun

The **BIG** Geographic Question

What roles do physical and cultural geography play in how a country changes?

From the article you learned why and how Japan was able to close its doors from 1630 to 1853. In the map skills lesson you saw how Japan's physical geography made it possible for the shoguns to isolate the country. Now find out how Japan was able to enter the twentieth century as a modern nation.

A. Review the article and construct a time line that chronicles events in Japan from 1603 to 1853. (Students' time lines might be similar to the following.)

1603	1630	1853
Emperor appoints Tokugawa Ieyasu shogun of Japan; Emperor loses power	Shogun closes Japan's doors to all foreigners; trade only with Dutch and Chinese	Commodore Perry arrives with warships; forces Japan to sign trade agreement

B. List some physical, economic, and political characteristics of Japan during its period of isolation from 1630 to 1853.

Physical	Economic	Political
few ports: Nagasaki, Edo	trade only with Dutch, Chinese	Shogun controls Japan, divides it into regions with Daimyos as heads; Emperor becomes a figurehead
		strict control on movement of people
		Catholic missionaries are expelled from Japan; Japanese Christians are persecuted

---

C. List some physical, economic, and political characteristics of Japan during the Meiji Restoration Period from 1868 to 1912.

Physical	Economic	Political
increased number of ports due to increased trade	increased industrialization	constitution is in place
	trade with West	Emperor's power restored
	students studying in Europe, United States	

D. Compare the attitudes of the Tokugawa Period to those of the Meiji Restoration Period concerning issues of trade, outside ideas, and foreigners.

	Trade	Outside Ideas	Foreigners
Tokugawa Period	little trade with West; mostly with Dutch, Chinese	Shogun fearful of rebellion, distrust of West	missionaries expelled, no sharing of ideas
Meiji Restoration Period	increased trade with West for industrialization	students attend schools in West, Westerners invited in to teach	foreigners welcome so ideas of modernization can be shared

E. Ieyasu, the first shogun of the Tokugawa Period, encouraged trade initially, but later he and other shoguns expelled traders for fear of losing their power. Today Japan is a major trading nation with great economic power. Explain how Ieyasu might respond to present-day Japan.

(Students' answers should include ideas about why Ieyasu closed Japan and should explain why Ieyasu might feel the same today, or why he might change his mind upon seeing how strong a world power Japan is now.)

A. Look at the political and physical map of south Asia and complete the following.

1. List the six countries shown.

Bangladesh	Nepal
Bhutan	Pakistan
India	Sri Lanka

2. Name the three large bodies of water that border south Asia.
Indian Ocean, Arabian Sea, and the Bay of Bengal

3. What are the three major mountain ranges that surround India?
Western Ghats, Eastern Ghats, and the Himalayas

4. What are the three main rivers in south Asia?
Ganges, Indus, and the Brahmaputra

B. Look for the precipitation and vegetation maps of India in the Almanac. Precipitation maps show the amount of rainfall an area receives. Vegetation maps show how plant life responds to climatic conditions. Complete the following.

1. Which parts of India receive the most rainfall?
the western coast of India and parts of northeast

2. What type of vegetation is found in the areas with the most rainfall?
broadleaf evergreen, broadleaf deciduous, and mixed deciduous and evergreen trees

3. Describe the general relationship between high levels of rainfall and vegetation.
Large amounts of rain produce thick vegetation.

C. Make connections among what you have learned about rainfall, vegetation, and physical features. Answer the following questions.

1. What kinds of physical features are located on the western coast and north-eastern coasts of India?
mountains

2. How do these physical features affect rainfall during the monsoon season?
Mountain slopes lift warm, moist air from the ocean to higher altitudes. The air cools and forms clouds that produce rain.

Lesson 14

ACTIVITY
Learn how the monsoon seasons affect human activity.

Living with Monsoons

The **BIG** Geographic Question

How do people adapt to major changes in the weather and climate of their environment?

From the article you learned how monsoons develop and where they occur. In the map skills lesson you learned about the connections among elevation, rainfall, and vegetation and how they relate to monsoons. Now see how the people of the region are affected by these changes in their physical environment.

A. Answer the following questions using information from the article and map skills lesson.

1. Describe the two monsoon seasons. How are they different from each other?

winter monsoon season: November–March; cool, dry air blows over land with occasional light rains

summer monsoon season: June–September; warm, moist air with warm temperatures and heavy rain

2. Explain how mountains help produce rain.

Mountain slopes lift warm, moist air from ocean to higher altitudes. The air cools and forms clouds that produce rain on the windward side of the mountains.

3. India is a country in south Asia that is greatly affected by monsoons. Why do you think this is so?

There are mountains on the north, east, and west boundaries of India. The east and west boundaries are also surrounded by water. The mountains and the water help produce rain.

B. Answer the following questions about how many people in south Asia make their living.

1. On what kind of activity do many people rely to make a living?
farming

2. What is the main crop produced in south Asia?
rice

3. How do monsoons affect the crops of farmers in south Asia?
When monsoons are on time and bring enough rain, farmers' crops will grow. Sometimes rains do not come or they arrive too late to be beneficial to the crops. If there is too much rain, crops will be destroyed. Also, villages are often damaged by flooding.

C. Imagine that you are a rice farmer living between the coast and the Western Ghats in southwest India. You have harvested your rice crop, and it is time to take it to market. However, the monsoons are still bringing heavy rains. More than five feet of rain has fallen. You need to sell your rice in order to earn money to buy items that your family needs.

Think about your 30-mile journey to the market. How will the heavy rains affect it? How will you transport your goods? How will you keep your rice dry? What other problems might you encounter?

Write a letter home to your family to let them know about your trip. Describe the journey and how the monsoon affected it.

(Students' letters should describe the difficulties of travel and daily life during the monsoon season.)

LESSON 15

A. Using the article and the map, answer the following questions.

1. What three seas surround Singapore, and to what two oceans does Singapore have access?

The South China Sea, Andaman Sea, and Java Sea surround Singapore. Singapore has access to the
Pacific and Indian oceans.

2. Why is Singapore important to East-West trade in terms of location?

Singapore is important to East-West trade because of its access to two major oceans and its free port
status. Singapore allows goods to be loaded or unloaded without having import taxes placed on them.
Singapore's location on major bodies of water in southeastern Asia make it very accessible to other
eastern countries as well as to western European and North American countries for trade.

B. Consider how other countries in the southeast Asia region are affected by their geography by answering the following questions.

1. What countries in the region are small and compact rather than scattered and sprawling?

Singapore, Brunei, and Taiwan

2. What countries in the region are made up of a series of sprawling islands?

Indonesia and the Philippines

3. What countries in the region are not compact but are relatively large?

Myanmar, Thailand, Laos, Vietnam, Cambodia, and Malaysia

C. Use what you learned from the map and your own ideas to answer these questions.

1. Do you think a long, spread-out nation (such as Vietnam) would be harder to unite than a small, compact nation (such as Singapore)? Why or why not?

A long, spread-out nation would probably be harder to govern because people would be dispersed
over a larger area, so transportation and communication would require more effort.

2. Do you think it would be easier to manage the economy of a small nation (such as Brunei) or a large nation (such as Indonesia)? Why?

The economy of a small nation would probably be easier to manage because the nation's management would be more centralized and its officials could more easily control business and industry.

LESSON 15

Lesson 15

ACTIVITY
Compare Singapore with other major economic powers in the region.

Four "Little Tigers"

The BIG Geographic Question
How has geography affected economic power in southeast Asia?

From the article you learned how geography played a major role in the economic development of Singapore. In the map skills lesson you explored whether a country's location and land area affect its management. Now find out how Singapore compares with the other three "Little Tigers"— Hong Kong, South Korea, and Taiwan.

A. The four "Little Tigers" are known for their economic power. Economic power is a country's ability to effectively produce (make), distribute (share), and consume (use) wealth, goods, and services. Use the Almanac to help you complete the chart below and get a picture of the economic power of the four "Little Tigers."

	Singapore	Hong Kong	South Korea	Taiwan
Value of Imports	$66.4 billion	$149.6 billion	$79.1 billion	$77.1 billion
Value of Exports	$61.5 billion	$145.1 billion	$80.9 billion	$85 billion
Major Trading Partners	U.S., Japan, Hong Kong, European Union	U.S., U.K., Japan, Germany, China, Taiwan	U.S., Japan	U.S., Hong Kong, Japan, Germany
Unemployment Rate	1.9%	2.3%	2.4%	1.6%
Hourly Wage	$5.25	$4.29	$5.53	$5.22

LESSON 15

B. Analyze the information on your chart and answer the following questions.

1. Which countries' export values are greater than their import value?

South Korea and Taiwan

2. Which countries' import and export values almost show a balance?

Singapore and Hong Kong

3. What do the low unemployment rate tell you about each of the "Little Tigers"?

Low rates show that each of the "Little Tigers" has a population that is primarily working and
suggests that the countries' economies are healthy.

C. Use the world map in the Almanac to identify the location of each of the four "Little Tigers." Then complete the following.

1. Describe the locations of each country.

a. Singapore: on the tip of Malaysia, access to Andaman, Java, and North China seas which connect
to the Indian and Pacific oceans

b. Hong Kong: on the southeast coast of China; access to the South China, Philippine, and East
China seas which connect to the Pacific Ocean

c. South Korea: a peninsula off the northeast coast of China; access to the Sea of Japan and the
East China Sea which connect to the Pacific Ocean

d. Taiwan: an island off the southeast coast of China; access to China via the Formosa Strait;
bordered on its east by the Philippine Sea which connects to the Pacific Ocean

2. Write a general explanation of how the countries' locations have helped their economic power.

All four countries' locations on major waterways have given them access to major East-West trade
routes. This has increased the countries' economic power.

D. Imagine you work for the Ministry of Economic Development for one of the "Little Tigers." You have been asked to create a brochure to encourage companies in the United States to invest. Use the article and other reference materials to prepare an outline of the brochure. Remember to use what you know about your country's geography.
(Students' outlines for brochure should include information on the country's port, government, industries, and economy as well as its location.)

LESSON 16

A. Use the time zone map to answer the following questions.

1. How many time zones are there in Australia?

three

2. If it is 10 P.M. in Sydney, what time is it in Perth?

8 P.M.

3. How many time zones are there in New Zealand?

one

4. If it is 10 P.M. in Sydney, what time is it in Wellington, New Zealand?

12 P.M.

The international date line is an imaginary line at 180° longitude that runs through the middle of the Pacific Ocean. By international agreement, it marks the spot where each new calendar day begins. Any time you cross the date line going from west to east, you move into the previous day (subtract one day). Anytime you cross the international date line going from east to west, you move into the following day (add one day). The time of day does not change. If you reach the date line at 4 P.M. on Tuesday coming from the east, it will be 4 P.M. Wednesday after you cross the line. If you cross back two hours later, it will be 6 P.M. on Tuesday.

B. With the previous information and the following scenario in mind, answer the questions below.

You are traveling from Western Samoa to Wellington, New Zealand. The trip takes four hours. You begin your trip at 7 A.M. on Saturday.

1. In which direction will you be traveling?

southwest

2. How many time zones will you cross during your trip?

one

3. Is New Zealand east or west of the international date line?

west

4. What day and time will you arrive in Wellington, New Zealand?

11 A.M. on Sunday

LESSON 16

Lesson 16

ACTIVITY
Determine whether the islands of the Pacific form a unified region.

The Sum of Island Parts

The BIG Geographic Question
What characteristics of an area establishes it as a unified region?

In the article you learned about the island groups of Oceania. In the map skills lesson you learned where the islands of the Pacific are located in relation to the international date line and how they fall into different time zones. Now determine whether the Pacific Islands form a physically, culturally, and politically unified region.

A region is an area of Earth having one or more common characteristics that are found throughout it. A region can be defined by human factors or by physical features. In each of the four regions of Oceania, the physical geography and history of the people is varied.

A. Choose two countries from one of the four regions of Oceania and compare them. Use the map and Almanac information to complete the chart. (Make sure students' answers reflect correct interpretations of the Almanac information and that islands have been chosen to represent the appropriate island group—Melanesia, Micronesia, Polynesia, or Australia.)

	Country Name	Country Name
Geographic Area		
Landforms		
Climate		
Crops		
Language		
Population		
Ethnic Groups		
Settlement History		
Government		

LESSON 16

B. Consider the characteristics of the various types of regions by completing the following questions.

1. What might be some of the characteristics of a physical region?

agriculture, climate, vegetation, and landforms

2. What might be some of the characteristics of a political region?

economy, military acquisition, government, zoning

3. What might be some of the characteristics of a cultural region?

language, religion, values, and nationality

C. Use the information you have collected to answer the following questions.

1. Do islands in the same region all have the same physical characteristics?

While some of the islands in each region may share some of the same physical characteristics, some
may be very different from each other. Some islands may be high islands, while others may be low
islands. Some are mountainous, while others are flat. The amount of rainfall may also vary considerably.

2. Do the Pacific Islands form a cultural region? Why or why not?

While each region may have similar cultures within islands, the Pacific Islands as a whole do not form
a cultural region. There is a great deal of cultural diversity among the Pacific Islanders.

3. Do the Pacific Islands form a politically united region? Why or why not?

No, the islands have many different types of government.

4. Can the area be identified as a physical region? Why or why not?

Yes, all the landforms are islands located in the Pacific Ocean and share a similar climate, except
for Australia because of its size.

A. To help you make this comparison, complete the following.

1. Draw the regional boundaries shown on Garreau's map onto the topographical region map. Use a colored pen or pencil such as red or blue that will show up well.

2. Now write in the names Garreau gave to those regions. In this way, you will be **superimposing** maps, or putting one map on top of another.

B. Use the article map to help you locate and label the following cities on the superimposed map. Then answer the questions below.

1. Label the city of Winnipeg, Canada. To which regions can you say Winnipeg belongs?

   a. Interior Lowlands _____

   b. the Breadbasket _____

2. Label the city of Toronto, Canada. To which regions can you say it belongs?

   a. Interior Lowlands _____

   b. the Foundry _____

3. Label Miami, Florida. To which regions does it belong?

   a. Gulf-Atlantic Coastal Plain _____

   b. the Islands _____

4. Label Los Angeles, California. To which regions does it belong?

   a. Pacific Mountains and Valleys _____

   b. MexAmerica _____

C. Review your answers for section B above. Draw some conclusions about the conclusions about the regions identified.

1. The names of the regions originally shown on the superimposed map describe what?

physical features of North America

2. The names of the regions from Joel Garreau's "nine nations" map describe what?

economic activity or common cultural characteristics

3. Now find a map in the Almanac showing different regions. What are the regions?

agricultural (crops) regions of the United States

Lesson 17

ACTIVITY
Choose capitals for the "Nine Nations of North America."

New Nations' Capitals

The BIG Geographic Question
Which characteristics should be considered in the selection of capital cities for various regions?

From the article you learned about Joel Garreau's "nine nations" of North America. The map skills lesson showed you an alternative way to regionalize North America. Now select a capital city for each of Garreau's regions.

A. Collect information about the "nine nations" from the article. Copy the below chart onto a sheet of paper and complete it. Use the Almanac to help you identify cities of North America. One has been done for you.

"Nation"	Cities Included	Primary Language(s)	Physical Features	Major Economic Activities
Ecotopia	Vancouver, San Francisco	English	mountains, forests, coast	logging, mining, recreation
Empty Quarter	Edmonton, Denver	English	mountains, forests, plains	mining, farming, ranching
Breadbasket	Winnipeg, Minneapolis	English	plains, prairies	farming, industry
Foundry	Ottawa, Chicago, New York City	English	prairies, hills	industry, farming
MexAmerica	Los Angeles, Mexico City	Spanish, English	mountains, desert	industry, farming, ranching
Dixie	Louisville, Dallas	English	mountains, warm climate	agriculture, tourism, hydroelectric dams
Quebec	Quebec, Montreal	French, English	forests, lakes, rivers	logging
New England	Boston, Halifax	English	hills, coastline	industry, farming, fishing
Islands	Miami, Havana	Spanish, English, French	largely islands, tropical climate	tourism, agriculture, industry

---

B. Imagine that each of these "nations" needs to choose a capital city. Use what you have learned about them to make an appropriate selection. Write your choices on the lines below, and support each choice with reasons why you selected it. For instance, you might choose Cleveland for the capital of The Foundry because it is centrally located and is an important manufacturing and shipping center on Lake Erie with access to the St. Lawrence Seaway and the Atlantic Ocean. (Students' answers should be supported by reasons that reflect facts about each region and the city selected.)

Ecotopia _____

The Empty Quarter _____

The Breadbasket _____

The Foundry _____

MexAmerica _____

Dixie _____

Quebec _____

New England _____

The Islands _____

A. Draw the following ocean routes between New York City and San Francisco on the map. (Make sure students routes are drawn according to the descriptions below.)

1. The shortest route going around Cape Horn at the southern tip of South America
2. The shortest route via the Panama Canal

B. Figure out the approximate distance and travel time for the routes you drew on the map. (Students' answers should be similar to the following.)

1. Use a piece or string and a ruler to measure the length of each route in inches.

   a. The route via Cape Horn is about ___15___ inches long.

   b. The route via the Panama Canal is about ___6___ inches long.

2. Now use the map scale to calculate the approximate number of miles in each route. (Hint: Multiply the number of inches in the route times the number of miles per inch to get the total length of the route.)

   a. The route via Cape Horn is about ___15,000___ miles long.

   b. The route via the Panama Canal is about ___6,000___ miles long.

3. The *Mauretania*, one of the fastest ocean-going steamships in the early 1900s, could travel about 745 miles per day. About how many days would it take the *Mauretania* to travel each route from New York City to San Francisco?

(Hint: Divide the length of route in miles by the number of miles per day to get the number of days needed to travel the route.)

   a. The voyage via Cape Horn would take about ___20___ days.

   b. The voyage via the Panama Canal would take about ___8___ days.

4. How many days would the ship save by using the Panama Canal?

12 days

5. Based on what you have learned from the above, write a couple of sentences explaining how the trip from New York City to San Francisco was affected by technological advances. Be sure to describe what those technological advances were.

The *Mauretania*, the fastest ocean going ship in the 1900s, decreased the number of days it took to travel from New York City to San Francisco around Cape Horn from 4 months to 20 days. The trip was also affected by the Panama Canal, which provided a shorter route that took even less time.

Lesson 18

ACTIVITY
Choose a good location for a new Central American canal.

Planning a New Canal

The BIG Geographic Question
What would be a good site for a new canal joining the Atlantic and Pacific oceans?

From the article you learned about both the advantages and disadvantages of the site for the Panama Canal. In the map skills lesson you saw how the Panama Canal changed the travel distance and time of trips by ship from the east to west coasts of North America. Now plan the site of a new canal in another location on the Central America isthmus.

A. Study the Central American map on page 116 in the Almanac. Think about the features of each country in Central America that might make it a good place or a poor place for a new canal to meet the world's shipping needs in the twenty-first century.

1. Which countries have the narrowest widths?

Panama and Costa Rica

2. Which countries have the narrowest mountain ranges?

Panama, Nicaragua, and Costa Rica

3. Why do you think Belize and El Salvador are not possibilities for building a new canal in Central America?

They are not possibilities because neither of these countries have coastlines on both the Atlantic and Pacific Oceans.

B. Use the Almanac and other research materials to find out more about the five countries that have coasts on both oceans. In the chart below list what you think would be the advantages and disadvantages of each country as a site for a new canal. What physical features do you see on the map that would help a country or put it at a disadvantage in the building of a canal? (Students' answers should list clear advantages and disadvantages supported by facts they can document.)

Country	Advantages	Disadvantages
Guatemala		
Honduras		
Nicaragua		
Costa Rica		
Panama		

C. Which country would you choose as the best site for a new canal? State your preference and support it with reasons.

_____

_____

_____

_____

McGRAW-HILL LEARNING MATERIALS
Offers a selection of workbooks to meet all your needs.

Look for all of these fine educational workbooks
in the McGraw-Hill Learning Materials SPECTRUM Series.
All workbooks meet school curriculum guidelines and correspond to
The McGraw-Hill Companies classroom textbooks.

SPECTRUM GEOGRAPHY – NEW FOR 1998!

Full-color, three-part lessons strengthen geography knowledge and map reading skills. Focusing on five geographic themes including location, place, human/environmental interaction, movement and regions. Over 150 pages. Glossary of geographical terms and answer key included.

TITLE	ISBN	PRICE
Grade 3, Communities	1-57768-153-3	$7.95
Grade 4, Regions	1-57768-154-1	$7.95
Grade 5, USA	1-57768-155-X	$7.95
Grade 6, World	1-57768-156-8	$7.95

SPECTRUM MATH

Features easy-to-follow instructions that give students a clear path to success. This series has comprehensive coverage of the basic skills, helping children to master math fundamentals. Over 150 pages. Answer key included.

TITLE	ISBN	PRICE
Grade 1	1-57768-111-8	$6.95
Grade 2	1-57768-112-6	$6.95
Grade 3	1-57768-113-4	$6.95
Grade 4	1-57768-114-2	$6.95
Grade 5	1-57768-115-0	$6.95
Grade 6	1-57768-116-9	$6.95
Grade 7	1-57768-117-7	$6.95
Grade 8	1-57768-118-5	$6.95

SPECTRUM PHONICS

Provides everything children need to build multiple skills in language. Focusing on phonics, structural analysis, and dictionary skills, this series also offers creative ideas for using phonics and word study skills in other language arts. Over 200 pages. Answer key included.

TITLE	ISBN	PRICE
Grade K	1-57768-120-7	$6.95
Grade 1	1-57768-121-5	$6.95
Grade 2	1-57768-122-3	$6.95
Grade 3	1-57768-123-1	$6.95
Grade 4	1-57768-124-X	$6.95
Grade 5	1-57768-125-8	$6.95
Grade 6	1-57768-126-6	$6.95

SPECTRUM READING

This full-color series creates an enjoyable reading environment, even for below-average readers. Each book contains captivating content, colorful characters, and compelling illustrations, so children are eager to find out what happens next. Over 150 pages. Answer key included.

TITLE	ISBN	PRICE
Grade K	1-57768-130-4	$6.95
Grade 1	1-57768-131-2	$6.95
Grade 2	1-57768-132-0	$6.95
Grade 3	1-57768-133-9	$6.95
Grade 4	1-57768-134-7	$6.95
Grade 5	1-57768-135-5	$6.95
Grade 6	1-57768-136-3	$6.95

SPECTRUM SPELLING – NEW FOR 1998!

This series links spelling to reading and writing and increases skills in words and meanings, consonant and vowel spellings and proofreading practice. Over 200 pages in full color. Speller dictionary and answer key included.

TITLE	ISBN	PRICE
Grade 1	1-57768-161-4	$7.95
Grade 2	1-57768-162-2	$7.95
Grade 3	1-57768-163-0	$7.95
Grade 4	1-57768-164-9	$7.95
Grade 5	1-57768-165-7	$7.95
Grade 6	1-57768-166-5	$7.95

SPECTRUM WRITING

Lessons focus on creative and expository writing using clearly stated objectives and pre-writing exercises. Eight essential reading skills are applied. Activities include main idea, sequence, comparison, detail, fact and opinion, cause and effect, and making a point. Over 130 pages. Answer key included.

TITLE	ISBN	PRICE
Grade 1	1-57768-141-X	$6.95
Grade 2	1-57768-142-8	$6.95
Grade 3	1-57768-143-6	$6.95
Grade 4	1-57768-144-4	$6.95
Grade 5	1-57768-145-2	$6.95
Grade 6	1-57768-146-0	$6.95
Grade 7	1-57768-147-9	$6.95
Grade 8	1-57768-148-7	$6.95

SPECTRUM TEST PREP from the Nation's #1 Testing Company

Prepares children to do their best on current editions of the five major standardized tests. Activities reinforce test-taking skills through examples, tips, practice and timed exercises. Subjects include reading, math and language. 150 pages. Answer key included.

TITLE	ISBN	PRICE
Grade 3	1-57768-103-7	$8.95
Grade 4	1-57768-104-5	$8.95
Grade 5	1-57768-105-3	$8.95
Grade 6	1-57768-106-1	$8.95
Grade 7	1-57768-107-X	$8.95
Grade 8	1-57768-108-8	$8.95

Look for these other fine educational series available from McGRAW-HILL LEARNING MATERIALS.

BASIC SKILLS CURRICULUM
A complete basic skills curriculum, a school year's worth of practice! This series reinforces necessary skills in the following categories: reading comprehension, vocabulary, grammar, writing, math applications, problem solving, test taking and more. Over 700 pages. Answer key included.

TITLE	ISBN	PRICE
Grade 3 – new for 1998!	1-57768-093-6	$19.95
Grade 4 – new for 1998!	1-57768-094-4	$19.95
Grade 5 – new for 1998!	1-57768-095-2	$19.95
Grade 6 – new for 1998!	1-57768-096-0	$19.95
Grade 7	1-57768-097-9	$19.95
Grade 8	1-57768-098-7	$19.95

BUILDING SKILLS MATH
Six basic skills practice books give children the reinforcement they need to master math concepts. Each single-skill lesson consists of a worked example as well as self-directing and self-correcting exercises. 48pages. Answer key included.

TITLE	ISBN	PRICE
Grade 3	1-57768-053-7	$2.49
Grade 4	1-57768-054-5	$2.49
Grade 5	1-57768-055-3	$2.49
Grade 6	1-57768-056-1	$2.49
Grade 7	1-57768-057-X	$2.49
Grade 8	1-57768-058-8	$2.49

BUILDING SKILLS READING
Children master eight crucial reading comprehension skills by working with true stories and exciting adventure tales. 48pages. Answer key included.

TITLE	ISBN	PRICE
Grade 3	1-57768-063-4	$2.49
Grade 4	1-57768-064-2	$2.49
Grade 5	1-57768-065-0	$2.49
Grade 6	1-57768-066-9	$2.49
Grade 7	1-57768-067-7	$2.49
Grade 8	1-57768-068-5	$2.49

BUILDING SKILLS PROBLEM SOLVING
These self-directed practice books help students master the most important step in math – how to think a problem through. Each workbook contains 20 lessons that teach specific problem solving skills including understanding the question, identifying extra information, and multi-step problems. 48pages. Answer key included.

TITLE	ISBN	PRICE
Grade 3	1-57768-073-1	$2.49
Grade 4	1-57768-074-X	$2.49
Grade 5	1-57768-075-8	$2.49
Grade 6	1-57768-076-6	$2.49
Grade 7	1-57768-077-4	$2.49
Grade 8	1-57768-078-2	$2.49

THE McGRAW-HILL
JUNIOR ACADEMIC™ WORKBOOK SERIES

An exciting new partnership between the world's #1 educational publisher and the world's premiere entertainment company brings the respective strengths and reputation of each great media company to the educational publishing arena. McGraw-Hill and Warner Bros. have partnered to provide high-quality educational materials in a fun and entertaining way.

For more than 110 years, school children have been exposed to McGraw-Hill educational products. This new educational workbook series addresses the educational needs of young children, ages three through eight, stimulating their love of learning in an entertaining way that features Warner Bros.' beloved Looney Tunes™ and Animaniacs™ cartoon characters.

The McGraw-Hill Junior Academic™ Workbook Series features twenty books – four books for five age groups including toddler, preschool, kindergarten, first grade and second grade. Each book has up to 80 pages of full-color lessons such as: colors, numbers, shapes and the alphabet for toddlers; and math, reading, phonics, thinking skills, and vocabulary for preschoolers through grade two.

This fun and educational workbook series will be available in bookstores, mass market retail outlets, teacher supply stores and children's specialty stores in summer 1998. Look for them at a store near you, and look for some serious fun!

TODDLER SERIES
32-page workbooks featuring the Baby Looney Tunes™

	ISBN	PRICE
My Colors Go 'Round	1-57768-208-4	$2.25
My 1, 2, 3's	1-57768-218-1	$2.25
My A, B, C's	1-57768-228-9	$2.25
My Ups & Downs	1-57768-238-6	$2.25

PRESCHOOL SERIES
80-page workbooks featuring the Looney Tunes™

	ISBN	PRICE
Math	1-57768-209-2	$2.99
Reading	1-57768-219-X	$2.99
Vowel Sounds	1-57768-229-7	$2.99
Sound Patterns	1-57768-239-4	$2.99

KINDERGARTEN SERIES
80-page workbooks featuring the Looney Tunes™

	ISBN	PRICE
Math	1-57768-200-9	$2.99
Reading	1-57768-210-6	$2.99
Phonics	1-57768-220-3	$2.99
Thinking Skills	1-57768-230-0	$2.99

GRADE 1 SERIES
80-page workbooks featuring the Animaniacs™

	ISBN	PRICE
Math	1-57768-201-7	$2.99
Reading	1-57768-211-4	$2.99
Phonics	1-57768-221-1	$2.99
Word Builders	1-57768-231-9	$2.99

GRADE 2 SERIES
80-page workbooks featuring the Animaniacs™

	ISBN	PRICE
Math	1-57768-202-5	$2.99
Reading	1-57768-212-2	$2.99
Phonics	1-57768-222-X	$2.99
Word Builders	1-57768-232-7	$2.99

SOFTWARE TITLES AVAILABLE FROM McGRAW-HILL HOME INTERACTIVE

The skills taught in school are now available at home! These titles are
now available in retail stores and teacher supply stores everywhere.
All titles meet school guidelines and are based on
The McGraw-Hill Companies classroom software titles.

MATH GRADES 1 & 2

These math programs are a great way to teach and reinforce skills used in everyday situations. Fun, friendly characters need help with their math skills. Everyone's friend, Nubby the stubby pencil, will help kids master the math in the Numbers Quiz show. Foggy McHammer, a carpenter, needs some help building his playhouse so that all the boards will fit together! Julio Bambino's kitchen antics will surely burn his pastries if you don't help him set the clock timer correctly! We can't forget Turbo Tomato, a fruit with a passion for adventure who needs help calculating his daredevil stunts.

Math Grades 1 & 2 use a tested, proven approach to reinforcing your child's math skills while keeping them intrigued with Nubby and his collection of crazy friends.

TITLE	ISBN	PRICE
Grade 1: Nubby's Quiz Show	1-57768-011-1	$19.95
Grade 2: Foggy McHammer's Treehouse	1-57768-012-X	$19.95

MISSION MASTERS™ MATH AND LANGUAGE ARTS

The Mission Masters™ -- Pauline, Rakeem, Mia, and T.J. – need your help. The Mission Masters™ are a team of young agents working for the Intelliforce Agency, a high level cooperative whose goal is to maintain order on our rather unruly planet. From within the agency's top secret Command Control Center, the agency's central computer, M5, has detected a threat… and guess what – you're the agent assigned to the mission!

MISSION MASTERS™ MATH GRADES 3, 4 & 5

This series of exciting activities encourages young mathematicians to challenge themselves and their math skills to overcome the perils of villains and other planetary threats. Skills reinforced include: analyzing and solving real world problems, estimation, measurements, geometry, whole numbers, fractions, graphs, and patterns.

TITLE	ISBN	PRICE
Grade 3: Mission Masters™ Defeat Dirty D!	1-57768-013-8	$29.95
Grade 4: Mission Masters™ Alien Encounter	1-57768-014-6	$29.95
Grade 5: Mission Masters™ Meet Mudflat Moe	1-57768-015-4	$29.95

MISSION MASTERS™ LANGUAGE ARTS GRADES 3, 4 & 5 – COMING IN 1998!

This new series invites children to apply their language skills to defeat unscrupulous characters and to overcome other earthly dangers. Skills reinforced include language mechanics and usage, punctuation, spelling, vocabulary, reading comprehension and creative writing.

TITLE	ISBN	PRICE
Grade 3: Mission Masters™ Feeding Frenzy	1-57768-023-5	$29.95
Grade 4: Mission Masters™ Network Nightmare	1-57768-024-3	$29.95
Grade 5: Mission Masters™ Mummy Mysteries	1-57768-025-1	$29.95

FAHRENHEITS' FABULOUS FORTUNE

Aunt and Uncle Fahrenheit have passed on and left behind an enormous fortune. They always believed that only the wise should be wealthy, and luckily for you, you're the smartest kid in the family. Now, you must prove your intelligence in order to be the rightful heir. Using the principles of physical science, master each of the challenges that they left behind in the abandoned mansion and you will earn digits to the security code that seals your treasure.

This fabulous physical science program introduces kids to the basics as they build skills in everything from data collection and analysis to focused subjects such as electricity and energy. Multi-step problem-solving activities encourage creativity and critical thinking while children enthusiastically accept the challenges in order to solve the mysteries of the mansion. Based on the #1 Physical Science Textbook from McGraw-Hill!

TITLE	ISBN	PRICE
Fahrenheit's Fabulous Fortune	1-57768-009-X	$29.95
Physical Science, Grades 8 & Up		

All titles for Windows 3.1™, Windows '95™, and Macintosh™.

Visit us on the Internet at
www.mhhi.com